# Byproducts from Agriculture and Fisheries: A Journey of Converting Waste to Wealth

# Byproducts from Agriculture and Fisheries: A Journey of Converting Waste to Wealth

**Saket Kushwaha**

Delve
PUBLISHING

www.delvepublishing.com

**Byproducts from Agriculture and Fisheries: A Journey of Converting Waste to Wealth**

*Saket Kushwaha*

**Delve Publishing**
224 Shoreacres Road
Burlington, ON L7L 2H2
Canada
www.delvepublishing.com
Email: orders@arclereducation.com



ISBN: 978-1-77407-614-9 (Hardcover)

# ABOUT THE AUTHOR

**Prof. Saket Kushwaha**, currently the Vice Chancellor of Rajiv Gandhi University is specialized in resource management and sustainable agriculture development. After his higher studies from Banaras Hindu University, he joined Abubakar Tafawa Balewa University (ATBU), Bauchi, Nigeria in 1993 and taught various courses on Agriculture Economics and Management at undergraduate, post graduate and Ph.D. level. Prof. Kushwaha rose to the rank of Professor at ATBU in the year 1999 and in 2006 he joined Bananas Hindu University (BHU), India as professor in agriculture economics and became the Vice Chancellor of Lalit Narayan Mithila University for one term 2014-2017. He has more than 100 publications in national and international journals of repute, supervised 24 Ph.D. students and authored 17 books / book chapters. Prof Kushwaha is life member of 10 Professional Bodies and sits in the panel of editorial boards. Worked extensively in the field of Zero Emission Research Initiatives (ZERI) propagating the mission of sustainable development under the aegis of "Waste is Wealth" concept. He is also the recipient of 17 national and international awards which includes award from Sulabh International Gold Medal in 2016 for sanitation management. 27 years of teaching, research, extension and community service experience with 23 years in administration. Handled over 10 national and International projects majorly funded by World Banks. Coordinated USAID project on Cowpea Research Support Programme (CRSP) in Nigeria for 10 years from 1996 to 2006. Working with NGOs and mega agriculture farms for Green Farm Planning.

# TABLE Of CONTENTS

# LIST OF FIGURES

# LIST OF TABLES

# LIST OF ABBREVIATIONS

| | |
|---|---|
| ACE | angiotensin in converting enzyme |
| ACTH | adrenocorticotropic hormone |
| ADR | aerobically digested residue |
| BGP | biogas plants |
| BIS | Bureau of Indian Standards |
| CBS | cotton burr and sticks |
| CF | crude fiber |
| CFP | common fisheries policy |
| CGB | cotton guayule bagasse |
| $CO_2$ | carbon dioxide |
| CP | crude protein |
| CS | cattle slurry |
| DHA | docosahexaenoic acid |
| DM | dry matter |
| DOC | controlled designation of origin |
| DRD | dried rumen digesta |
| EAF | ecosystem approach to fisheries |
| EPA | eicosatetraenoic acid |
| EU | European Union |
| FDA | Food and Drug Administration |
| FPC | fish protein concentrate |
| GCP | Guayule entire plant |
| GES | good environmental status |
| GFCM | General Fisheries Commission for the Mediterranean |
| HDF | human dermal fibroblasts |
| HPLC | high-performance liquid chromatography |
| ICCAT | International Commission for the Conservation of Atlantic Tuna |
| IEBPs | inedible by-products |

| | |
|---|---|
| LME | liposoluble marc extract |
| LPS | lipopolysaccharide |
| MIN | mineral fertilizer process |
| MITF | microphthalmia-associated transcription factor |
| MPAs | marine protected areas |
| NO | nitric oxide |
| NOS | nitric oxide synthase |
| NPK | nitrogen, phosphorous, potassium |
| OVW | oil vegetation waters |
| OWF | oak wood fiber filler |
| PAC | powdered activated carbon |
| PCBs | polychlorinated biphenyls |
| PGE2 | prostaglandin E2 |
| PHP | powdered pistachio hull |
| PMS | pulp mill sludge |
| PUFA | polyunsaturated fatty acids |
| TOS | theory of sampling |
| TSE | transmissible spongiform encephalopathy |
| TTX | tetrodotoxin |
| UHPLC | ultra-high-performance liquid chromatographic |
| USDA | United States Department of Agriculture |
| UV-C | ultraviolet C |
| WHC | water holding capacity |
| WHO | World Health Organization |
| ZTB | zone di tutela biological |

# PREFACE

Meeting the dietary requirements of an ever-growing and affluent universal population is a primary challenge for human safety and economic security. Moreover, there is also a bulk human population (nearly 1 billion) who are suffering from malnutrition due to food scarcity. Many analysts assess that the production of food needs to approximately double over the impending several years. However, there is a potential to alleviate food scarcity by intelligently using the by-products from agriculture, livestock, and fisheries. Recent tendencies of consuming food crops for fulfilling fuel requirements (e.g., corn ethanol) along with replacement of fuel crops with food crops have resulted in amplified food scarcity. Therefore, there is a need to utilize the by-products from the agriculture and seafood industry while continually cultivating these two industries. Moreover, the effect of climate change on agriculture, livestock, and fisheries also needs to be monitored and coped with. The potential capability of fishery and agricultural systems to inhibit climate change also needs to be explored. There is a wide range of products which can be manufactured using by-products from agricultural and fishery wastes. The end products manufactured from fishery and agricultural by-products include biofuels, building materials, cosmetics, pharmaceuticals, etc.

This book on by-products from Agricultural and Fishery wastes explains the technologies and processes involved in this procedure of reconstitution which focuses on waste management, recycling, and climate control. The book essentially contains a detailed discussion about environmental-friendly approaches to develop value-added products in the domains of food security Internationally recognized author has contributed insightful and engaging chapters on a broad selection of plant and animal by-products, discoursing the real-world business of byproduct retrieval within the dynamic contexts of fluctuating socioeconomic distresses.

This book contains eight chapters. The first two chapters deal with the introduction of agricultural and fishery wastes. Chapter 1 focuses on generic fundamentals of wastes produced from agriculture, livestock, and fisheries. On the other hand, Chapter 2 deals with characterization and treatments for agricultural wastes. Chapter 3 discusses the different uses of animal by-products in food systems. Moreover, the chapter also highlights some of the concerns associated to the use of inedible by-products. Chapter 4 essentially focuses on the medicinal uses of animal by-products.

Plant-based by-products can be used in various industries. Chapter 5 illustrates the use of plant wastes and by-products in the cosmetics industry. Chapter 6 comprehensively discusses the potential uses of plant-based-By-products for compost manufacturing. Chapter 7 introduces the readers with the fundamentals of fisheries and their by-products. Finally, Chapter 8 discusses the innovations, ecology, and economy of fisheries and their potential by-products.

With its mixture of broader commentary and practical instruction, this book offers crucial insight and knowledge to all students, researchers, teachers, policymakers, and professionals interested in agriculture, food science, environmental science, seafood industry, and various other interdisciplinary fields. This book can be used as a ready reference for all the readers who intend to get a deep insight into agricultural and fishery by-products.

**—Author**

# 1

# *An Introduction to Agricultural and Fishery Wastes*

## CONTENTS

## 1.1. INTRODUCTION

The description of waste material produced from numerous fishery and farming operations is broadly referred by the term agricultural and fishery waste. These activities encompass dairy farming, livestock cultivation, wild fish harvest and aquaculture, poultry egg production, and also fruits and vegetable farming. By-products are co-produced by few of the produce inferred from the above-listed activities and their conversion into shelf-stable forms; they are contemplated either unfit or unsalable for human devouring. The further classification of the waste is non-natural and natural (organic) types (Ashworth and Azevedo, 2009). The non-natural types involve plastic packaging waste, fertilizer absence, and other agrochemicals like pesticides released into the habitat, fishing fleet parts such as batteries and tires and farm vehicles, discarded pesticide containers, and others like oil containers and oil residue, outmoded machinery, and many more. The natural waste comprises of residues coming from the processing and growing of untreated agricultural and fishery products (e.g., fish, livestock, dairy products, fruits, vegetables, and eggs), and food and agricultural industry refuse (slaughterhouses, fish processing plants, and manure and wastewater from farms).

There are several forms in which agricultural and fishery wastes crop up such as liquids, solids, or slurries; and establish a significant fraction of the world-wide agricultural and fishery intake. The agricultural and fishery regions produce these sufficient wastes that have been diversely estimated at above 30% of the overall agricultural and fishery fertility. They are perpetually high in nutrient and moisture contents, which drive them to tremendously perishable and incline to degradation by endogenous enzymes and microorganisms degrade. Hence, if they are not transformed into the value-added, balanced forms, then they highly affect environmental pollution capacity. Despite their unfavorable environmental effects, particular biohazard threats are also posed by agricultural and fishery waste in the manner of waterborne diseases, animal-borne diseases, and pathogens (like fungi, bacteria, viruses, and molds). For instance, respiratory problems (like occupational asthma) can be caused by the exposition of farmworkers to farm dust that originated from waste such as animal manure. Injurious health effects like disorders of the central nervous system like restlessness, convulsions, and memory loss; eye irritations; likewise, incessant diseases like cancers, reproductive, and assorted developmental disorders can be drawn out by the disclosure to farm pesticides as well as other agrochemicals (Magauzi et al., 2011). However, agricultural, and fishery wastes are prosperous sources of valuable nutrients and other elements that can be rehabilitated and invested in rewarding use. Consequently, there can have both beneficial and adverse effects of agricultural and fishery wastes, and this book will approach the advantages and disadvantages of these wastes in the current environment (Figure 1.1)

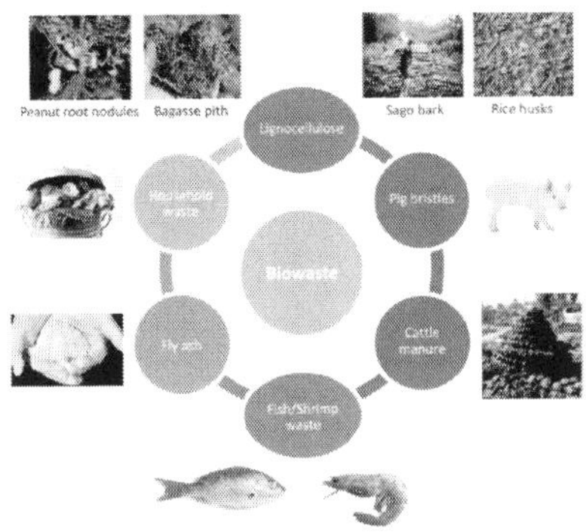


**Figure 1.1:** Classification of different types of biowastes (including agricultural, livestock, and marine wastes).

*Source: https://pubs.rsc.org/en/content/articlelanding/2019/cs/c8cs00543e.*

At present, the generously formed waste materials are chiefly consumed as biogas, animal feed, fertilizer, pigments, and cosmetics (Jayathilakan et al., 2012). The following sections present a detailed discussion on waste materials related to numerous industrial zones. In ancient times, a large number of wastes was utilized as feed, fertilizer, dumped in landfills, and burnt outright or water bodies. However, the earlier methods of controlling agriculture and fishery waste are not any more acceptable for the modern community due to present rigid environmental safety standards and expanding food production. This book will address different fishery and plant by-products, livestock, and a few contemporary waste utilization strategies. Note that this chapter contains the preliminary introduction only, and discuss an abstract overview of waste produced by the agricultural and fishery zones, and some of the manners in which they are consumed.

## 1.2. LIVESTOCK WASTE

Livestock may be expressed in broad or narrow terms. It may be explained more broadly to involve caged wild animals like elk, deer, monkey, rabbit, moose, buffalo, etc.; or narrowly to cover tamed animals like pigs, goats, cattle, sheep, horses, donkeys, camels, llamas, and poultry (for example, turkeys, chicken, ducks, guinea fowl, geese, squabs). Either way, as food for human devouring, milk, and meat are produced by livestock mammals, whereas eggs and meat are produced by poultry. Feather, leather, wool, and fur are the other commodities deduced from livestock lifting. Labor is also provided by any of these animals, i.e., the camel, buffalo, horse, and donkey; whereas others also represent pets (e.g., cats, birds, dogs, horses, and guinea pigs) (Figure 1.2).

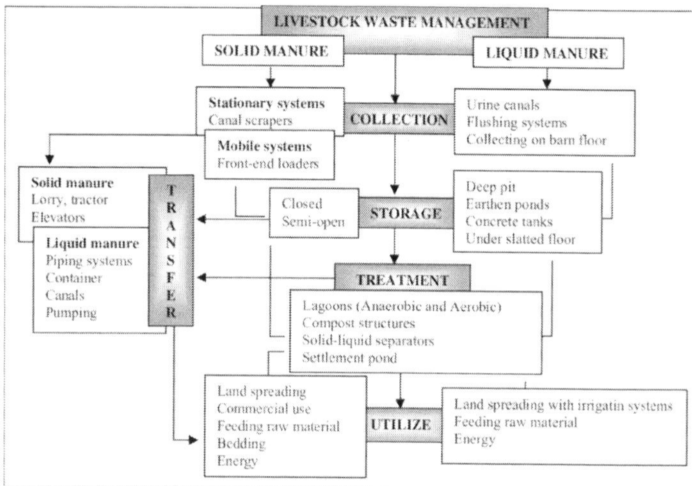


**Figure 1.2:** Different modules of waste management for livestock wastes.

*Source: https://www.researchgate.net/figure/Components-of-livestock-waste-management-system_fig1_325286137.*

There are two leading stages of livestock production for milk, eggs, and meat; especially, slaughtering, and raising the animals. Throughout the raising step, animals grow to a definite sought body weight in readiness for practice as human food; and in the course of the second phase, the animal is slayed, wiped, and sliced into portions according to the requirement of consumers. Just the same to be anticipated, when the animal is full of life, animal manure creates a dominant agricultural waste, while the extensive agricultural waste of the deceased animal is formed by body parts and tissues, and also wastewater arising out of processing operations. The many popular wastes and livestock by-products are dealt with in this section. Postharvest disposes of taken out of the slaughterhouse incorporate factors such as fatty tissue, blood, hide, heads, viscera, tail, feet, and specific internal organs (e.g., gizzard, brain, kidney, and heart), wastewater, and manure. It is observed that in some traditions, a few of the slaughterhouse dumps mentioned before are coveted, expensive delicacies, for example, the brain, blood, heart, intestine, stomach, tongue, tail, and kidney.

### 1.2.1. Manure

The massive enclosed animal feeding operations in consideration of their milk, eggs, or meat raise most of the livestock, e.g., pigs, cattle, sheep, poultry, horses, goats, rabbits, turkeys, and chicken, etc. A substantial quantity of animal manure is produced in the process by these animals which are also named animal dung. Manure is a combination of urine and raw excrement, along with hair, soil, and waste feed. Divergent animals have diverse qualities of manures. For example, there are high N and C contents in cow manure; high N and K contents in chicken and sheep manure; aside from that there are low N and K contents in pig manure. Horses are basically herbivores that's why they cannot digest grains; so, horse manure can include weed seeds and grass (Zhang and Schroder, 2014). Hence, there is the varying formation of manure, additionally, it is affected by determinants such as the composition of the food given to the animal, type of animal (i.e., non-ruminant vs ruminant), the animals' atmosphere, age, and nutritional situation of the animal (Figure 1.3)

**Figure 1.3:** Animal manure waste stacked together.

*Source: https://www.lcdri.com/applications/animal-manure-waste/.*

Manure management results in animals' manure creation, transferring the from the spot of generation to collection store, treatments (by biological, chemical, and physical methods) at the storehouse to minimize their potential of pollution, and retaining it in storehouse facilities for the following usage. The livestock manure is consumed in several ways including as the source of energy for the formulation of biomethane; or by gasification, pyrolysis, or combustion to create electricity or heat; as animal feed ingredient; as bedding stuff for animals; as fertilizer and organic matter for grass and crop production; or as mulch. For instance, in the plant required nutrients such as P, N, Ca, K, Fe, Mg, and Zn, manure is high (Combs et al., 1998). Therefore, organic matter is used to furnished soils by using animal manure being a fertilizer and raised their physicochemical abilities to keep plants green and healthy; and also aids the soil by regulating soil pH, growing nutrient and water uptake and loosening soil, and draining of nitrates towards the soil and lessening runoff (Ogejo, 2015).

### 1.2.2. Blood and Plasma

In the existent animal, the blood supply in the role of the fluid that transfer nutrients and $O_2$ to body tissues and cells, and at the same time transmitting waste materials from these similar tissues and cells. Its color is dark red because of the myoglobin, pigments, and hemoglobin, whose paramount atom, i.e., $Fe^{2+}$, empowers the uptake and transmission of gases, e.g., $CO_2$ and $O_2$. It is approximated that 3 to 4% of the weight of the animal's body is possessed by blood (Jayathilakan et al., 2012). Moreover, blood taken from livestock is high in heme iron and protein and is used as a nutrient in some civilizations.

As an example, some European countries used them in preparing products such as blood pudding and animal sausages; and in some Asian countries, it is used in the preparation of blood cakes and blood curds (Jayathilakan et al., 2012; Ghost, 2001). From the blood, plasma is also consumed as a stabilizer, an emulsifier, as a nutritional supplement in foods, colorant, or as a foaming agent in baked freight (Silva and Silvestre, 2003).

Blood also comprises$\alpha_2$-macroglobulin which is the general protease inhibitor that is utilized to hold back the texture depreciation of the muscle food induced by endogenous proteases (Ashie and Simpson 1996, 1994; Sareevoravitkul et al., 1996). Blood from livestock as plant fuel, and as animal food due to its immense mineral and vitamin contents, is as well offered to non-food consumption, e.g., water and fat cover by the generosity of its transglutaminase element (blood factor XIII) (Silva and Silvestre, 2003). The plasma from the blood is more than that utilized as an additive for the preparation of blood agar and as a media of tissue fashion (Kurbanoglu and Kurbanoglu, 2004) (Figure 1.4).

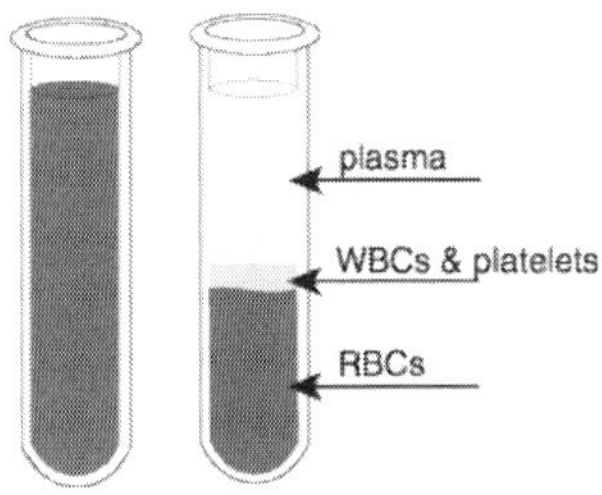


**Figure 1.4:** Characteristic composition of animal blood containing red blood cells (RBCs), white blood cells (WBCs), platelets, and plasma.

*Source: https://thefactfactor.com/facts/pure_science/biology/human-biology/blood-plasma/4537/.*

### 1.2.3. Skins

It is evaluated that approximate of 4 to 11% of the load of the alive animal is accounted for the hides or skins of livestock (except for poultry) (Benjakul et al., 2009). Conventionally, the skins or hides were manipulated as a point of supply of leather and relevant goods (bags, clothing, belts, automobile seats, shoes, etc.), and also glue, gelatin, cosmetics, sausage casings. Hides or skins engaged for such utilizations must be endogenous enzymes to prevent the breakdown of the tissues and curing with salt to eject microorganisms, or rapidly dealt with air-drying.

The skin has high collagen which is the triple-helical structural protein. Gelatin is deduced from collagen by enzymatic or chemical hydrolysis (Benjakul et al., 2012). Alteration to gelatin contains the primary elimination of non-collagenous protein material and after that the treatment of heat and hydrolysis of the heat-denatured collagen through to gelatin (Figure 1.5).

**Figure 1.5**: Animal skins extracted from sacrificial or dead animals.

*Source: https://www.aliexpress.com/item/32463804586.html.*

There are numerous industrial applications of gelatin. It is applied as a fat confiner in meat commodities or as an emulsifier in the food industry. It is additionally served as a stabilizer in yogurt, ice creams, cream pies, and jellies. Gelatin is conjointly used as a compounding or coating agent in ointments, in tablets and cosmetic productions and for manufacturing glue. There is film-forming characteristic of gelatin, that's why it is applicable in injury curing for dressing burns, and in the production of biodegradable plastic films, and as skin transplant.

### 1.2.4. Bones

It is supposed that around 10 to 20% of an animal carcass is composed of bones (Jayathilakan et al., 2012; MPI 2018). Bones contain a tremendous protein content (roughly 11 to 16%), which is broadly chondroitin-rich proteoglycans and collagen. The amount of mineralization of collagen in bones is greatly superior to that of the skin, which is the basis for the strongly inflexible structure of skins versus bones. Hence, similar to skins, bones are also applicable to produce gelatin. For the domestic purpose, bones are expended to cook soups to draw out the flavors and marrow for human devouring. The insoluble bone material might be included into animal food as a source of nourishment by the reason of their high content of vital vitamins (e.g., $B_{12}$), amino acids, and minerals (e.g., Ca); orare smashed and worked as food nutritive fortifier or as pet feed for poultry and dogs. Also, bones are opted as fuel in the shape of bone-charcoal acquired after dry distillation, or as a fertilizer due to its high phosphorous content. Slaughterhouse bones are used by others include as knife handles, buttons, as a component of porcelain, and in soap making (Figure 1.6)

**Figure 1.6:** Dump site for animal bones and wastes.

*Source: https://stock.adobe.com/search?k=%22animal+bones%22.*

### 1.2.5. Internal Organs

The internal organs of livestock consist of the heart, brain, kidneys, gizzards, lungs, liver, rumen, pancreas, stomachs, abomasum, tongue, intestines, udder, testes, spleen, and the uterus. They are expressed as delicacies, and are eminently cherish as food in few of the cultures, and high in nutrients, and have obvious textures and flavors. Many Asian and African countries used the internal organs of livestock as traditional medicine. A few of the glands like the pituitary, pancreas, thyroid, parathyroid, corpus luteum, ovary, kidney, and others, secrete hormones that serve to handle reproductive problems in women (e.g., estrogen and progesterone from pig ovaries) and to control metabolism. Additional organs, i.e., the hypothalamus, spinal cord, and brain have high cholesterol which can be extracted and treated as an emulsifier for the synthesis of vitamin $D_3$ and in cosmetics (Ejike and Emmanuel, 2009). The lungs and liver secrete heparin that is used as a blood thinner to avert blood clotting and an anticoagulant. The pancreas is an origin of the insulin, glucagon, and hormone. Insulin conduct glucose metabolism, as well as is used to medicate diabetes of type 1. In contrast, glucagon is used to raise blood sugar levels, and to medicate low blood sugar or insulin surplus. The pancreas also delivers a host of digestive enzymes, e.g., the lipases, and serine proteases (trypsin and chymotrypsin), used in wound healing; and in the textiles, food, and detergent industries, etc. The intestines of calves and sheep are served as sausage casings, or to manufacture surgical sutures.

### 1.2.6. Livestock Fat

Animal fats are collected from the (i) slaughter, (ii) processing, and (iii) packaging of pigs, cows, chicken, sheep, and other animals. Animal fats may be depleted as human food trust in the quality. The animal fat type appropriate for human usage is particularly high-quality lard usually extracted from wet-rendering from healthy pigs, and high-quality tallow generally deduced by wet-rendering from healthy sheep or cattle (Figure 1.7).

**Figure 1.7:** Bulk livestock fat stored in a pot.

*Source: https://educalingo.com/en/dic-en/animal-fat.*

There is direct use of high-quality lard in human feed; nonetheless, to bleach it and to freshen it is the on-going fashion before its use in food. Tallow and Lard were used typically for the purpose of deep-frying. But, considering public health this approach of use is decreasing. At present, lard, and tallow are consumed as an emulsifier in sausages and alternate emulsified foods, and for producing shortening and margarine (Ghotra et al., 2002). Uneatable greases and tallow, and low-quality lard, are used as industrial fats, as an energy source, or in animal feeds, and are formed from the animals generally by dry rendering.

### 1.2.7. Feathers

Discards or by-products of poultry production are feathers. The primary component of feathers is the structural protein keratin. There is a crystalline structure of feathers that are stable, sturdy, and light in weight. A large surface area is associated with them that makes them able to ingest more moisture as compared to wool or cellulose. Feathers are fit for decorative materials, e.g., feather fans, costume accessories, fishing flies, masks, earrings, ornaments, and as flowers. Additionally, feathers are servicing to manufacture a range of products, i.e., as adsorbent and padding for diapers, mattress, and pillow stuffing, padding, and insulation for fabrics, fertilizer, and high protein animal feed (feather feast).

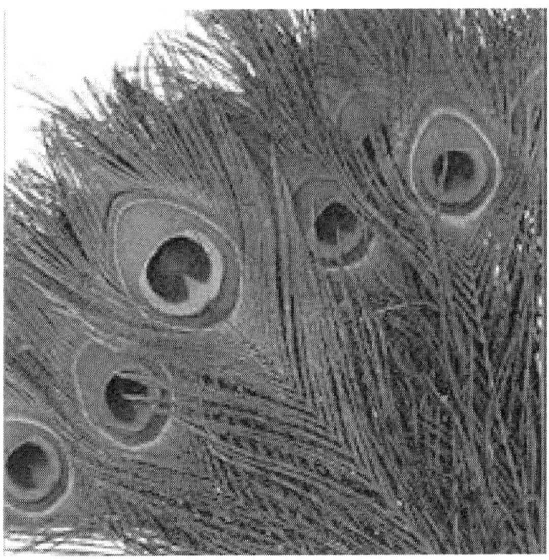

**Figure 1.8:** Peacock feathers are extensively used for decorative purposes.

*Source: https://english.newstracklive.com/news/peacock-feather-use-can-improve-your-bad-luck-condition-know-how-to-use-it-here-sc77-nu-1011247-1.html.*

### 1.2.8. Egg Shells

Female birds and alternate species produce eggs to treat as food or to spread their species. Poultry eggs expected for the use of human beings come about the unfertilized eggs given by adult female birds that are chickens, or to an insignificant extent, alternate species such as goose, duck, quail.

The outermost covering of the poultry eggs is shells. The shell made up of around 9.5% of the overall egg weight and is basically constituted of calcium carbonate (98.4%) with some calcium phosphate (0.8%), magnesium carbonate (0.8%), and organic matter (0.4%) chiefly protein. For the exchange of gases, about 7000 to 17000 pore canals per egg are present. For the prevention of microbial contamination, the pore canals that are full of protein fibers, are closed by a cuticle layer comprised of keratin. According to the breed, the color of the eggshell is between the dark brown and chalky white and is because of pigments called ooporphins that are scattered on the shell surface (Figure 1.9).

**Figure 1.9:** Cracked egg-shells can be used for many purposes which include medicinal uses, pest repelling and soil nutritioning.

*Source: https://thisismygarden.com/2020/01/use-egg-shells/.*

Below the shell, two shell membranes, a thicker outer one and an inner one, lie. Jointly, they create an approximate of 0.015 mm layer. The membranes are constituted of keratin and their involved thread network indicates a constitutional tactic of security as opposed to microbial contamination.

There are several uses of eggshells. For example, like mineral-rich powder, the shells are mashed and cleaned that are worked as a nutritional additive to protect confectioneries, beverages, and baked goods. In the production of low phosphate feed, the crushed shells have been examined also by virtue of its high Ca content, for individuals agonizing from kidney collapse. It also acts as a fertilizer, and due to its high $CaCO_3$ content used for fertilizing to subdue truly acidic pH soils.

Their scraping nature at the time of crushing arise a separate unpopular application of eggshells. Scattering crushed shells over the plants hold bash off the plants. The abrasive quality of crushed eggshells is also oppressed for unblocking drains and clogged sinks and used as a facial scrub. The eggshell membrane raised under the shell is comprised of collagen that is applicable in the cosmetics, biomedical, and numerous other industries.

## 1.3. FISHERY WASTE

Fishery gathering and processing produce a huge amount of waste material that is predicted to range from 35–80% of the overall animal. The wastes encompass processing waste coming from desk inning, de-heading, evisceration, deboning, trimming, washing, filleting operations, and peeling, and by-catch, and dismissed from fish harvesting. The solid waste material involves bones, heads, skin, cartilages, fins, and tails, scales, hepatopancreas, carapace, viscera, shells, pyloric ceca, legs, and tails (Ghaly et al., 2013). The wastewater or soluble waste is acquired from defrosting, bleeding, cooking, cleaning, and canning operations; and has dissolved pigments, high protein contents, and salts. These wastes all contain beneficial nutrients and elements such as proteins, enzymes, pigments, fat, carotenoid pigments, chitins, anti-freeze proteins, flavorants, etc., that may be used to create high value-added products or to preserve the hike of living organisms (Ramalingam et al., 2014; Simpson 2007, 2000) (Figure 1.10).

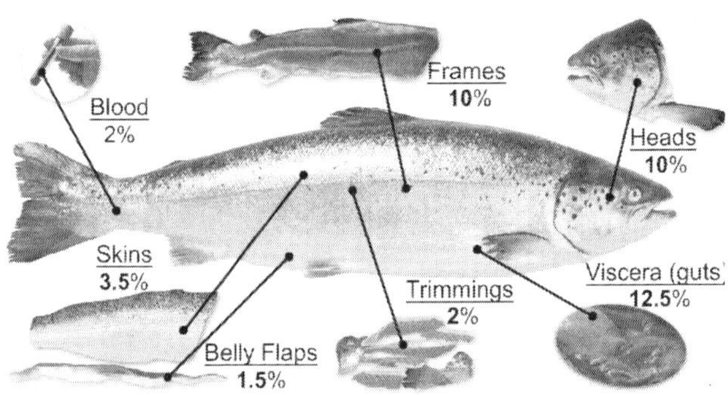


**Figure 1.10:** Typical structure and components of a fish.

*Source: https://thefishsite.com/articles/dont-bypass-the-value-of-aquaculture-by-products.*

### 1.3.1. Fish Viscera

Fish viscera is comprising of factors like the intestines, stomachs, pancreas, hepatopancreas, and pyloric ceca. They are the worthwhile point of supply of digestive enzymes such as carbohydrases (like lactase, pancreatic amylase, sucrase, and maltase), proteases, and peptidases (for example, trypsins, pepsins, carboxy peptidases, chymotrypsins, elastase, erepsin, and cathepsins), nucleases (for instance ribonuclease and deoxyribonuclease), and lipases (such as pancreatic lipase, gastric lipase, sterol esterase, and phospholipase) (Poonsin et al., 2018; Aryee et al., 2007; Noriega-Rodríguez et al., 2009).

It is richly embossed by Gildberg et al. (2000), digestive enzymes coming out of fishery waste carry excess uses. Following are few of these uses: for the removal of connective tissue corresponding with fish roe, fish proteases from crab hepatopancreas are used in the construction of caviar to upgrade yield; for the eradication of skate wings, fish scales, and fish skins using a combination of carbohydrase and proteases arising out of fishery waste; the supplementation of fish fermentation, and the use of cod trypsin for the readiness of carotenoproteins originating at crustacean waste (Poonsin et al., 2018; Gildberg et al., 2000; Haard and Simpson, 1994). More uses of fishery waste enzymes are for enhancing the kind of feed given to farmed fish, and for the manufacturing of fish sauce and cheddar cheese.

### 1.3.2. Other Fishery Wastes

There is high protein content in fish heads, except for that they are usually disposed of. On the other hand, they can be used in the production of fish protein hydrolysates, fish meal, feed supplement, fertilizer, or fish paste. Fish skins are utilized being the sources of gelatin that are applicable in the pharmaceutical, cosmetic, and food industries as they have high fat and high collagen contents. Leather products are also manufactured by fish skins, e.g., belts, shoes, wallets, purses, bags, and others. As scraps, a big chunk of fish meat associated with the bones is left by fish filleting that can be revived and turned into products like fish sauce that all have high mineral content, or fish burgers, fish minces, fish dumplings, fish hydrolysates, and fish sticks. They are mixed with surimi, canned fish and a kind of other foodstuffs, e.g., pizzas, pasta, soups.

Fishery waste has a great content of protein that designates them wonderful nitrogen sources, hence they can also act as a fertilizer. Calcium carbonate as an extra mineral is also added to the soil by the shellfish shells. The fish food for animal feed can be obtained by the transformation of those shellfish and fish that may be rejected by fish harvesting and by-catch. This is achieved through homogenization of the waste, after that pressing to evacuate liquid to retrieve the residue, on milled and evaporated to generate a dark brown powder. A by-product of fishmeal manufacturing is fish oil. Once the solids happen to pressed and milled, the oil is separated from the aqueous stage by the centrifugation of the liquid homogenate (Figure 1.11).

**Figure 1.11**: Fish oil is extracted from fish wastes which can be used for various purpose.

*Source: http://activeauthorities. com/nature-made-fish-oil-review/.*

The usage of fish oil is as feedstock for the production of biodiesel, and in paints (Aryee et al., 2011). A massive amount of solid organic scrap is also involved in wastewater coming from the fishery region, which can be fermented into biomethane for the usage purpose as a domestic heating fuel (Achinas et al., 2017). The extensive amount of wastewater arising out of the fishery industry or effluents contains notable amounts of P and N, that cause them beneficial as fertilizer.

## 1.4. VEGETABLE AND FRUIT WASTE

The collection and processing of vegetables and fruits produce a large number of wastes that consists of any of the unused from numerous processing operations, and also the harvesting dismissed. The kinds and amount of waste coming out of the diverse commodities varies as there are various species of nuts, vegetables, and fruits. For instance, the waste produced by slicing of papaya constitute of seeds (~6.5%), peels (~8.5%), pulp waste (~32%);apple alteration into juice gives around 11% of pulp and seeds as waste; the waste generated from the processing of mangoes with the final product is about 42% in form of seeds, peels, unusable pulp; nearly 16% of peels is co-produced by peeling of mandarins with the peeled finished product; whereas pineapple processing returns about 40% of waste (like pulp, core, tops) with the completed product (~48%), and around 14% of peels;. Those nuts that are produced with commercial quantities covers Brazil nuts, almonds, hazelnuts, chestnut, cashew, macadamia, pistachio, walnuts, and pecan. Examples of commercial quantities producing vegetables include Brussel sprouts, basil, cabbage, chives, carrots, cucumbers, onions, lettuce, parsley, mustard, peas, corn, potatoes, Swiss chard, spinach, tomato. The fruits cropped for commercial purposes are avocado, pineapples, cantaloupe, bananas, coconut, citrus fruits, apples, raspberries, grapes, olives, mangoes, coffee, pears, papaya, peaches, pomegranate, plums, strawberries, and others. The discarded waste materials generated by vegetables and fruits are commonly about 20 to 30% according to Schieber et al. (2001) and Ajila et al. (2010, 2007). The processing operations that set up the waste consist of canning, production of pasta and sauces and vegetable paste, production of non-alcoholic and alcoholic beverages, manufacturing of jellies and jam, drawing out of vegetable oils, and others. Some of the waste coming from these materials involves peels, leaves, rinds, pulp, pomace, roots, seeds, stems, skin, stalks, shells, tubers, vines, and stones (Panouille et al., 2007) (Figure 1.12).

**Figure 1.12:** Image showing a bulk of vegetable and fruit waste.

*Source: https://inhabitat.com/uk-report-finds-up-to-40-of-all-fruit-and-vegetables-are-wasted-because-of-ugliness/.*

### 1.4.1. Dietary Fiber

Waste materials drained from vegetables and fruits are abundant in dietary fiber, the indigestible fraction of plant meals that are thought to be important for health.

The bowel movement can be improved by dietary fiber in order to relieve or avert constipation. It defeats the danger of non-communicable diseases such as diabetes, obesity, and cardiovascular diseases and assists in weight management as it is not absorbed or digested.

Dietary fiber exists in fruit waste like onions, red grape, and apple pomace, mango, and apple peels, and many more (Yan and Kerr, 2013; Li et al., 2014).

### 1.4.2. Enzymes

An enzyme is obtained from the sources of the waste generated from the processing and harvesting of vegetables and fruits.

The enzymes normally found in plant materials are amylases, proteases, celluloses, lipases, peroxidases, polyphenol oxidizes, and lipoxygenases. Long-chain polypeptides and proteins are decomposed by the proteases into amino acids and low molecular weight peptides.

Commonly known plant proteases contain bromelain (inferred from the pineapple fruit and pineapple stem), ficin (obtained from the latex of the fig tree), and the papaya latex enzymes (i.e., chymopapain, papain, and protease III).

Plant lipases (also raised in seeds) acts to hydrolyze triglycerides into glycerol and free fatty acids.

Plant amylases chiefly contains β-amylases (established in seeds) which transforms polysaccharides into disaccharides. Cellulase (an enzyme complex made up of endo-1,4-β-d-glucanase, β-d-glucosidase, exo-1,4-β-glucanase) and lipoxygenases (acquired from seeds such that soybeans) are responsible for production of hydroperoxides and catalyze the extension of $O_2$ to polyunsaturated fatty acid molecules.

Arrange of higher plants have launched the celluloses, e.g., tomato, avocado, soybean, pepper, barley, and strawberry, and they evolve glyosidic bonds into fiber (Porta and Rocha-Sosa, 2002;

Ferrarese et al., 1995; Kemmerer and Tucker, 1994; Hoy et al., 1981) (Figure 1.13).

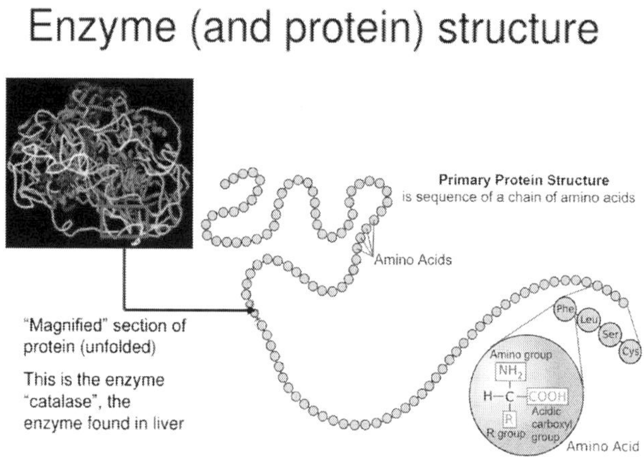


**Figure 1.13:** Structural representation of enzyme and protein.

*Source: https://www.slideserve.com/channing-melendez/enzyme-structure-and-function.*

For the production of microbial enzymes, vegetable, and fruit wastes have been adopted too as substrates. For instance, various microorganisms like *Aspergillus oryzae*, *Aspergillus niger*, *Bacillus subtilis*, and *Rhizopus oryzae* used to amylase production of banana waste and uphold its growth (Said et al., 2014). *Bacillus licheniformis,* and *B. subtilis* used palm kernel waste in the form of substrates for cellulose production by the way of solid-state fermentation (Norsalwani and Norulaini, 2012). A consolidation of citrus and bran hybrid (i.e., kinnow) waste, was utilized in the production of cellulose enzymes with the help of *Trichoderma reesei* (Oberoi et al., 2010). *Aspergillus flavus* and *Aspergillus niger* produced invertase through lactose, fruit peel, or sucrose in form of carbon sources (Mehta and Duhan, 2014). For the production of pectinases, several alternate vegetable and fruit wastes (such as orange, lemon peels, and pineapple) have been worked as carbon sources at the hand of *Penicillium chrysogenum* and *A. niger* (Okafor et al., 2010). Xylanases, tannases, and proteases are the other enzymes that are utilized in the food industry and that have been generated using vegetables and fruits waste materials with microorganisms. Solid-state fermentation produces many more with the help of microorganisms *A. oryzae*, *Penicillium atramentosum,* and *A. niger* on waste coming from palm kernel, pomegranate, tamarind, and leaves from several plants (Varadharajan et al., 2016; Kumar et al., 2007; Krishna, 2005). For the production of food thickeners, pigments, and vegetable oils, and for making the clear image of non-alcoholic and alcoholic beverages, Xylanases, tannases, and proteases are used broadly in food processing (Banerjee et al., 2005). Plant enzymes are used in other food industry such as papain, proteases, ficin, bromelain, along with their utilization to tenderize meats. For the production of starch syrup, fruit juices, chocolate cakes, moist cakes, and also used in a variety of processes like preparation of digestive aids, baking, and brewing, amylases are extensively practiced in the food industries (Laufenberg et al., 2009). For the extraction of natural pigments, food flavor compounds, polyphenolic compounds, and essential oils, the enzyme cellulose is used via plant food waste such as potato peels, grape pomace, and banana peels. For the explanation of liquor formed from vegetables and fruits, celluloses are used too, to boost their stability, yields, and visual allure to consumers. There is a large usage of enzyme invertase in the food industry for the production of jam, confectionery, and candies, and for the conversion of sucrose to invert sugar (Panda et al., 2016). The food and beverage industries apply pectinases to aid clarification, extraction, and concentration of the final products. There are also non-food uses of vegetable and fruit waste enzymes. For example, human intervertebral disc degeneration has investigated by chymopapain (Wardlaw, 2016), and in research works rheumatoid arthritis has investigated (Figure 1.14).

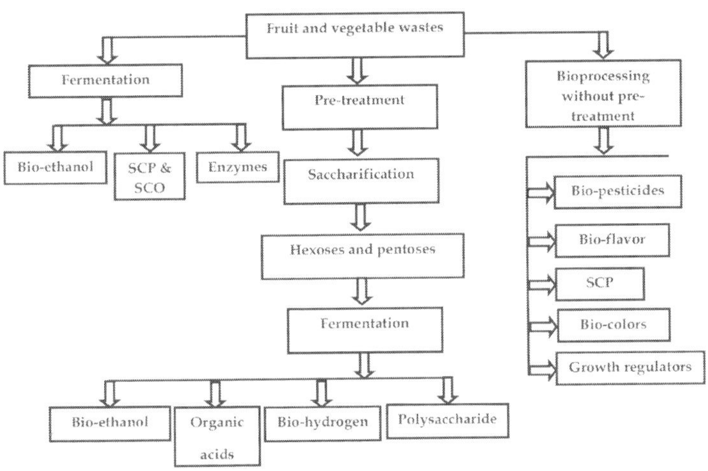


**Figure 1.14:** Flowchart for processing of vegetable and fruit waste to produce different end-products.

*Source: https://www.researchgate.net/figure/Bioprocessing-of-fruit-and-vegetable-wastes-into-potential-biocommodities_fig1_316798759.*

### 1.4.3. Polyphenolic Compounds

More useful biomolecules including polyphenolic compounds also raised in these wastes. Examples of such compounds are anthocyanins, flavanoids, lignans, stilbenes, coumarins, lignins, tannins, xanthones, and bioflavonoids (Balasundram et al., 2006), and they appear generously in assorted nuts, vegetable, and fruits waste like the peels, leaves, seeds, and rinds. For instance, banana leaves contain peonidin, petunidin, and cyanidin; apple leaves have phenolics such as chlorogenic acid, quercetin, epicatechin, and catechin; grape seeds carry procyanidins; red beet includes *p*-coumaric acids, and ferulic; mango seeds involve gallotannins and gallates; seeds peels and citrus encompass hesperidin, eriocitrin, and naringin; kiwifruit peels comprise of caffeic acid, and *p*-coumaric acid; garlic incorporates *p*-coumaric acid, caffeic acid, and ferulic acid; carrot pomace covers $\beta$- and $\alpha$-carotenes; tomato skins contain lycopene; pomegranate waste consists of ellagitannins, anthocyanins, gallagic acid, and punicalagin; and potatoes cover gallic acid, caffeic acid, and chlorogenic acid (Balasundram et al., 2006; Friedman, 1997; Turrini et al., 2016). The bioactive compounds accompanying a host of biological operations are known as polyphenolics, which accumulate essential health profit to consumers. Some carry antimicrobial impact, some possesses antioxidants characteristics, few others exhibit enzyme inhibitory impression, whereas others act as metal chelators. Other influence phenolic compounds show contain anti-inflammatory and anti-cancer and cardio-protective impact, cholesterol lowering, between others.

### 1.4.4. Flavor and Aroma Compounds

In various regions, flavor, and aroma compounds are essential elements for uses involving confectioneries, beverages, and food, pharmaceuticals, cosmetics, and health industries. Flavor and aroma compounds in the beverage and food industry only, comprise above a quarter of the world-wide food preservative market. Nowadays, the production of a lot of the flavoring compounds is done by chemical synthesis or by eradication from natural unprocessed materials (i.e., animals, microorganisms, or plants). Nonetheless, by regulatory preferences and in response to consumer, the trends of the current market are moving toward goods that can be categorized just as natural. Still, the problem occurs on the production of natural flavors using extraction from plants on the grounds of probable co-extraction of unacceptable parts in the untreated materials, least yields, or exhausting extraction processes. Hence, bioconversion or microbial biosynthesis is the better way to achieve the flavor and aroma compounds production. Because of the production of vegetable and fruit wastes in abundant amounts, and the up-gradation of most of them in nutrients like proteins, minerals, and carbohydrates; they can treat as substrates to assist development and aroma/flavor compounds composition by microorganisms. With the help of vegetable and fruit wastes being substrates, for the production of flavor and aroma compounds, large-scale studies have used various microorganisms (Longo and Sanromán, 2006). Such microorganisms can be *B. subtilis*, *A. niger*, *Kluyveromyces marxianus*, *Ceratocystis fimbriata*, *Lactobacillus amylophilus*, *Lactobacillus acidophilus*, *Pediococcus pentosaceus*, *Lactobacillus paracasei*, *Zygosaccharomyces rouxii*, and *Rhizopus oryzae* (Longo and Sanromán, 2006; Soares et al., 2000; Christen et al., 2000; Bramorski et al., 1998). While microorganisms have used to support the expansion and aroma/flavor production of the wastes, and the waste include apple pomace, amaranth waste, cocoa sweatings, and cocoa husks, cassava bagasse, coffee husks, coconut fat, licorice root wastes, garlic skins, palm bran, maize cobs, soybean curd residues, and soybean hulls, rice hulls, tobacco waste, and sugarcane bagasse (Figure 1.15).

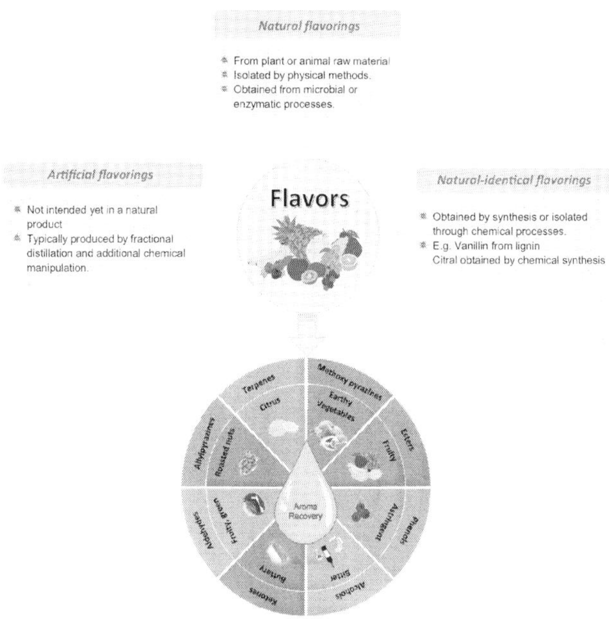


**Figure 1.15:** Sources of different flavors (natural and artificial).

*Source: https://link.springer.com/article/10.1007/s12393-017-9172-8.*

Numerous flavor active compounds like butyric acid, acetaldehyde, ethanol, diacetyl, isobutyl acetate, ethyl acetate, ethyl-3-hexanoate, isoamyl acetate, lactic acid, and 6-pentyl-a-pyrone (6-PP), 3-methylbutanol, pyrazines (e.g., tetra methylpyrazine, and 2,5-dimethylpyrazine), furanones, and vanillin are being produced by the operation of the microorganisms on the above-mentioned substrates. Peculiar flavors are being connected to a few of such compounds. For example, the dairy aroma is being related to lactic acid, diacetyl, and butyric acid;6-pentyl-a-pyrone (6-PP) is connected with coconut aroma; geranyl acetate and methyl butyrate pass on fruity aroma; pineapple aroma is provoked by methyl butanoate; isoamyl acetate extracts banana aroma; benzaldehyde participates in almond flavor; whereas vanillin is associated with vanilla flavor. Flavor and aroma constituents deduced from such wastes are utilized in the cosmetic, food, and pharmaceutical industries (Castilho et al., 2000).

### 1.4.5. Fruit and Vegetable Wastes as Livestock Feed

The extensive quantity of animal food proteins is provided by livestock in the shape of milk, meat, feed, and eggs. A large number of feed materials are required for livestock production that are to be nourished to the animals, moreover, livestock production should be secure, quickly available, replenish able, and economical reasonable. A huge amount, roughly yearly basis estimate that is millions of tons of wastes is generated from all over the world coming from the production and operation of nuts, vegetable, and fruit (Wadhwa and Bakshi, 2013). One of the reasons for environmental pollution is that most of the already stated waste is disposed of in rivers or landfills. Although, they are tremendous in minerals (e.g., K, Co, Ca, Mo, Mn, P, Na, Zn, and S), nutrients, and other valuable components (e.g., phenolics, carotenoid pigments, and many more), that could be settled to beneficial usage as nourishment for livestock (Figure 1.16).

**Figure 1.16:** Fruit and vegetable wastes are commonly used as feed for livestock.

*Source: https://about.sprouts.com/sustainability/environment/.*

Waste arising out of apple harvesting and operating, i.e., the pomace, is one of the examples of the wastes that bring to feed for livestock. Pomace in a variety of patterns is contained in the food of poultry and cattle (Ghoreishi et al., 2007). Plantain and banana production generate wastes consisting of damaged fruits, rejects, leaves, peels, pseudo stems, and stalks that were given in several forms to livestock. For instance, banana, and plantain wastes were ensiled along with more waste material like rice bran, grass, wheat straw, molasses, and others, and served as feed for cows, buffaloes, sheep, goats, poultry, and rabbits (Wadhwa and Bakshi, 2013; Khattab et al., 2000). The pomace and the culls are involved in tomato harvesting and operation. Tomato waste as feed is consumed in numerous forms for cattle, buffaloes, pigs, goats, and poultry (Bakshi et al., 2014; Ventura et al., 2009; Sethi, 2012; Sayed and Abdel-Azeem, 2009). For the feeds of goats, cows, and pigs, sugar beet waste was assimilated (Wadhwa and Bakshi, 2013). Other vegetables and fruits waste used in the act of livestock feed encompass carrots, cabbage, chickpeas, cauliflower, grapes, citrus fruits, potatoes, mangoes, squash, and radish. Several feeds for livestock have incorporated such wastes like horses, cows, sheep, rabbits, poultry, and pigs and were strongly endorsed by mentioned animals.

## 1.5. WASTE WATER

The universal leading users of water are the fishery and agricultural zones. Water is needed for livestock rearing, aquaculture, vegetable, and fruits farming, and growth by entire life forms (covering bacteria, worms, insects, etc.). This water is obtained from the freshwater point of supplies like lakes, wells, and rivers and from rainfall. Nevertheless, the use of freshwater sources like rivers becomes essential due to the seasonal essence of rainfall and nasty climatic situations as drought, to achieve rainfall for the purpose of irrigation and fish rearing, by all of the others. Freshwater is desired as well for managing and operating of fishery and agricultural products. The amount of water needed for such activities is overwhelming and lays strong pressure in contact with freshwater resources. Nonetheless, an excessive volume of wastewater is also co-produced by such activities that can be dealt with and reused to ease the pressure ahead of freshwater resources. The wastewater usage is connected to both profits and risks (Toze, 2006). The profits of usage of wastewater are as follows: for irrigation, it lessens the expense of energy for pumping or extracting groundwater resources; it makes proper operation of wastewater happen and its reuse relieve pressure ahead of freshwater resources markedly; and to minimize eutrophication of water bodies, and to preserve costs on fertilizer, it generally rebound available macronutrients (K, P, and Na) and micronutrients (Mg, Ca, Mg, B, Mn, Zn, or Fe) back to the soil (Liu and Haynes, 2011; Matheyarasu et al., 2016). There can be biological and chemical nature of risks in agriculture with unprocessed wastewater use. The biological risks with wastewater usage in corporate viruses (e.g., *Caliciviruses*, *Adenoviruses*, *Reoviruses*, *Rotavirus,* and *Noroviruses*);various bacteria (e.g., *Shigella*, *Escherichia coli*, Vibrio, and *Campylobacter*, *Salmonella*, *Arcobacte*r, *Helicobacter*, *Clostridium perfringens*, *Mycobacterium ulcerans,* and *Staphylococcus aureus*);helminths (e.g., *Ancylostoma duodenale, Ascaris lumbricoides*, *Strongloides stercoralis*, *Taenia saginata,* and *Trichuris trachiura*); *Schistosoma spp.* (also known as *blood-flukes*);and protozoans (e.g., *Giardia intestinalis*, *Entamoeba histolytica,* and *Cryptosporidium parvum*). Biological hazards can have destructive outcomes in the sense of waterborne ailments (such as cholera, diarrhea, gastroenteritis), conjunctivitis, typhoid, and respiratory tract infections (Mara et al., 2007). Alternate diseases include dysentery, salmonellosis, and Legionnaire's disease (*Legionella pneumophila*). Chemical risks are in the kind of polychlorinated biphenyls (PCBs) (e.g., phenoclor, clophen, arochlor), synthetic pesticides (e.g., herbicides, insecticides, and fungicides like propoxur, acephate, diazinon, malathion, and metaldehyde), heavy metals (e.g., Be, As, Pb, Hg, and Cd), and hydrocarbons (e.g., phenanthrene, pyrenes, furans, and dioxins).

## REFERENCES


1. Achinas, S., Achinas, V., & Euverink, G. J. W., (2017). A technological overview of biogas production from bio-waste. *Green Chemical Engineering-Review Engineering, 3(*3), 299–307.
2. Ajila, C. M., Aalami, M., Leelavathi, K., & Rao, U. P., (2010). Mango peel powder, a potential source of antioxidant and dietary fiber in macaroni preparations. *Innovative Food Science and Emerging Technologies, 11*, 219–224.
3. Ajila, C. M., Bhat, S. G., & Rao, U. P., (2007). Valuable components of raw and ripe peels from two Indian mango varieties. *Food Chemistry, 102*, 1006–1011.
4. Aryee, A. N. A., Simpson, B. K., & Villalonga, S. R., (2007). Lipase fraction from the viscera of grey mullet (*Mugil cephalus*). Isolation, partial purification and some biochemical characteristics. *Enzyme and Microbial Technology, 40*, 394–402.
5. Aryee, A. N. A., Simpson, B. K., Cue, R. I., & Phillip, L. E., (2011). Enzymatic transesterification of fats and oils from animal discards to fatty acid ethyl esters for potential fuel use. *Biomass Bioenergy, 35*, 4149–4157.
6. Ashie, I. N. A., & Simpson, B. K., (1994). Effect of hydrostatic pressure on alpha macroglobulin and selected proteases. *Journal of Food Biochemistry, 18(*6), 377–391.
7. Ashie, I. N. A., & Simpson, B. K., (1996). $\alpha_2$-Macroglobulin inhibition of endogenous proteases in fish muscle. *Journal of Food Science, 61(*2), 357–361.
8. Ashworth, G. S., & Azevedo, P., (2009). In: Ashworth, G. S., & Azevedo, P., (eds.), *Agricultural Wastes*. Nova Science Publishers.
9. Bakshi, M. P. S., Kaur, J., & Wadhwa, M., (2014). Nutritional evaluation of sun-dried tomato pomace as livestock feed. *Indian Journal of Animal Nutrition, 84(*5), 71–75.
10. Balasundram, N., Sundram, K., & Samman, S., (2006). Phenolic compounds in plants and agri-industrial by-products, antioxidant activity, occurrence, and potential uses. *Food Chemistry, 99*, 191–203.
11. Banerjee, R., Mukherjee, G., & Patra, K. C., (2005). Microbial transformation of tannin-rich substrate to gallic acid through co-culture method. *Bioresource Technology, 6*, 949–953.
12. Benjakul, S., Kittiphattanabawon, P., & Regenstein, J. M., (2012). Fish

gelatin. In: Simpson, B. K., Nollet, L. M. L., Toldra, F., et al., (eds.), *Food Biochemistry and Food Processing*, (Vol. 1, 388–405). Ames, IA, USA, Wiley.

13. Benjakul, S., Oungbho, K., Visessanguan, W., et al., (2009). Characteristics of gelatin from the skins of bigeye snapper, *Priacanthus tayenus* and *Priacanthus macracanthus*. *Food Chem., 116*(5), 445–451. https://www.sciencedirect.com/science/article/abs/pii/S0308814609002489.
14. Bramorski, A., Christen, P., Ramirez, M., et al., (1998). Production of volatile compounds by the edible fungus *Rhizopus oryzae* during solid-state cultivation on tropical agro-industrial substrates. *Biotechnology Letters, 20*, 359–362.
15. Castilho, L. R., Medronho, R. A., & Alves, T. L., (2000). Production and extraction of pectinases obtained by solid state fermentation of agro-industrial residues with *Aspergillus niger*. *Bioresource Technology, 71*, 45–50.
16. Christen, P., Bramorski, A., Revah, S., & Soccol, C. R., (2000). Characterization of volatile compounds produced by *Rhizopus* strains grown on agro-industrial solid wastes. *Bioresource Technology, 71*, 211–215.
17. Combs, S. M., Peters, J. B., & Zhang, L. S., (1998). Manure, trace element content and impact on soil management. *New Horizons Soil Science, 7*, 10–19. Madison, WI, Univ. Wisconsin http://www.uwex.edu/ces/forage/wfc/proceedings2001/micronutrient_status_of_manure.htm (accessed on 30 May, 2020).
18. Ejike, C. E., & Emmanuel, T. N., (2009). Cholesterol in kidneys of cow. *African Journal of Biochemistry Research, 3(*4), 95–97.
19. Ferrarese, L., Trainotti, L., Moretto, P., et al., (1995). Differential ethylene-inducible expression of cellulase in pepper plants. *Plant Molecular Biology, 29*, 735–747.
20. Friedman, M., (1997). Chemistry, biochemistry, and dietary role of potato polyphenols: A review. *Journal of Agricultural and Food Chemistry, 45*, 1523–1540.
21. Ghaly, A. E., Ramakrishnan, V. V., Brooks, M. S., et al. (2013). Fish processing wastes as a potential source of proteins, amino acids and oils, a critical review. *Journal of Microbial and Biochemical Technology* 5, 107–129.

22. Ghoreishi, S. F., Pirmohammadi, R., & Yansari, A. T., (2007). Effect of ensiled apple pomace on milk yield, milk composition and dry matter intake of Holstein cows. *Journal of Animal and Veterinary Advances* 6, 1074–1078.

23. Ghost, R., (2001). Fractionating of biological macromolecules using carrier phase ultra-filtration. *Biotechnology and Bioengineering, 74*, 1–11.

24. Ghotra, B. S., Dyal, S. D., & Narine, S. S., (2002). Lipid shortenings: A review. *Food Research International, 35(*2), 1015–1048.

25. Gildberg, A., Simpson, B. K., & Haard, N. F., (2000). Uses of enzymes from marine organisms. In: Haard, N. F., &. Simpson, B. K., (eds.), *Seafood Enzymes: Utilization and Influence on Postharvest Seafood Quality (*Vol. 1(22), pp. 619–640). New York. NY, Marcel Dekker.

26. Haard, N. F., & Simpson, B. K., (1994). Proteases from aquatic organisms and their uses in the seafood industry. In: Martin, A. M., (ed.), *Fisheries Processing, Biotechnological Applications* (Vol. 1(6), pp. 132–153). Boston, MA, Springer.

27. Hoy, J. L., Macauley, B. J., & Fischer, G. B., (1981). Celluloses of plant and microbial origin in germinating barley. *Journal of the Institute of Brewing, 87*, 77–80.

28. Jayathilakan, K., Sultana, K., Radhakrishna, K., & Bawa, A. S., (2012). Utilization of by-products and waste materials from meat, poultry and fish processing industries: A review. *Journal of Food Science and Technology, 49(*3), 278–293.

29. Kemmerer, E. C., & Tucker, M. L., (1994). Comparative study of celluloses associated with adventitious root initiation, apical buds, and leaf, flower, and pod abscission zones in soybean. *Plant Physiology, 104*, 557–562.

30. Khattab, H. M., Kholif, A. M., El-Alamy, H. A., et al. (2000). Ensiled banana wastes with molasses or whey for lactating buffaloes during early lactation. *Asian Australian Journal of Animal Sciences, 13*, 619–624.

31. Krishna, C., (2005). Solid-state fermentation systems: An overview. *Critical Reviews in Biotechnology, 25*, 1–30.

32. Kumar, R., Sharma, J., & Singh, R., (2007). Production of tannase from *Aspergillus ruber* under solid-state fermentation using jamun (*Syzygium cumini*) leaves. *Microbiological Research, 162*, 384–390.

33. Kurbanoglu, E. B., & Kurbanoglu, N. I., (2004). Utilization as peptone for glycerol production of ram horn waste with a new process. *Energy Conversion and Management, 45(*2), 225–234.
34. Laufenberg, G., Schulze, N., & Waldron, K., (2009). A modular strategy for processing of fruit and vegetable wastes into value-added products. In: Waldron, K. W., (ed.), *Handbook of Waste Management and Co-product Recovery in Food Processing* (Vol. 1, pp. 286–353). New York, NY, Woodhead Publishing Limited.
35. Li, X., He, X., Lv, Y., & He, Q., (2014). Extraction and functional properties of water-soluble dietary fiber from apple pomace. *Journal of Food Process Engineering, 37(*3), 293–298.
36. Liu, Y., & Haynes, R., (2011). Origin, nature, and treatment of effluents from dairy and meat processing factories and the effects of their irrigation on the quality of agricultural soils. *Critical Reviews in Environmental Science and Technology, 41*, 1531–1599.
37. Longo, M. A., & Sanromán, M. A., (2006). Production of food aroma compounds, microbial and enzymatic methodologies. *Food Technology and Biotechnology, 44(*3), 335–353.
38. Magauzi, R., Mabaera, B., Rusakaniko, S., et al., (2011). Health effects of agrochemicals among farmworkers in commercial farms of Kwekwe district, Zimbabwe. *The Pan African Medical Journal, 9*, 20–26.
39. Mara, D., Sleigh, P., Blumenthal, U., & Carr, R., (2007). Health risks in wastewater irrigation, comparing estimates from quantitative microbial risk analyses and epidemiological studies. *Journal of Water and Health, 5*, 39–50.
40. Matheyarasu, R., Seshadri, B., Bolan, N. S., & Naidu, R., (2016). Abattoir wastewater irrigation increases the availability of nutrients and influences on plant growth and development. *Water, Air, and Soil Pollution, 8*, 227–253.
41. Mehta, K., & Duhan, J. S., (2014). Production of invertase from *Aspergillus niger* using fruit peel waste as a substrate. *International Journal of Pharmacy and Biological Sciences, 5*, 353–360.
42. MPI. (2018). *Meat Processing Insight/Animal Bones Protein Market Research.* http://www.meat-machinery.com/meat-processing-insight/ html (accessed on 30 May, 2020).
43. Noriega-Rodríguez, J. A., Gámez-Meza, N., Alanis-Villa, A., et al., (2009). Extraction and fractionation of lipolytic enzyme from viscera

of Monterey sardine (*Sardinops sagax caerulea*). *International Journal of Food Science and Technology, 44(*6), 1223–1228.

44. Norsalwani, T. T., & Norulaini, N. N., (2012). Utilization of lignocellulosic wastes as a carbon source for the production of bacterial celluloses under solid state fermentation. *International Journal of Environmental Science and Development, 3*, 136–140.
45. Oberoi, H. S., Chavan, Y., Bansal, S., & Dhillon, G. S., (2010). Production of celluloses through solid state fermentation using kinnow pulp as a major substrate. *Food Technology, 3*, 528–536.
46. Ogejo, J. A., (2015). Manure production and characteristics. *Animal Manure Management.* (Vol. 1, pp. 1–44) http://articles.extension.org/pages/15375/manure-production-and-characteristics (accessed on 30 May, 2020).
47. Okafor, U. A., Okochi, V. I., Chinedu, S. N., et al. (2010). Pectinolytic activity of wild-type filamentous fungi fermented on agro-wastes. *African Journal of Microbiology Research, 4*, 2729–2734.
48. Panda, S. K., Mishra, S. S., Kayitesi, E., & Ray, R. C., (2016). Microbial-processing of fruit and vegetable wastes for production of vital enzymes and organic acids, biotechnology and scopes. *Environmental Research, 146*, 161–172.
49. Panouille, M., Ralet, M. C., Bonnin, E., & Thibault, J. F., (2007). Recovery and reuse of trimmings and pulps from fruit and vegetable processing. In: Waldron, K. W., (ed.), *Handbook of Waste Management and Co-product Recovery in Food Processing* (Vol. 1, pp. 417–447). Cambridge, UK, Woodhead Publishing Limited.
50. Poonsin, T., Simpson, B. K., Benjakul, S., et al., (2018). Carotenoprotein from Pacific white shrimp (*Litopenaeus vannamei*) shells extracted using trypsin from albacore tuna (*Thunnus alalunga*) spleen, antioxidant activity and its potential in model systems. *Journal of Food Biochemistry, 42(*2), 42, e12462. https://onlinelibrary.wiley.com/doi/abs/10.1111/jfbc.12462.
51. Porta, H., & Rocha-Sosa, M., (2002). Plant lipoxygenases. Physiological and molecular features. *Plant Physiology, 130(*1), 15–21.
52. Ramalingam, V., Thirunavukkarasu, N., Chandy, N., & Rajaram, R., (2014). Proximate composition of trash fishes and their utilization as organic amendment for plant growth. *Journal of the Marine Biological Association of India, 56(*2), 1–29. http://mbai.org.in/uploads1/

manuscripts/56-2-art2431769324.pdf.

53. Said, A., Leila, A., Kaouther, D., & Sadia, B., (2014). Date wastes as substrate for the production of α-amylase and invertase. *Iranian Journal of Biotechnology, 12*, 41–49.
54. Sareevoravitkul, R., Simpson, B. K., & Ramaswamy, H., (1996). Effects of crude $\alpha_2$macroglobulin on properties of bluefish (*Pomatomus saltatrix*) gels prepared by high hydrostatic pressure and heat treatment. *Journal of Food Biochemistry, 20(*6), 49–63.
55. Sayed, A. N., & Abdel-Azeem, A. M., (2009). Evaluation of dried tomato pomace as feedstuff in the diets of growing rabbit. *International Journal for Agro Veterinary and Medical Sciences, 3*, 13–18.
56. Schieber, A., Stintzing, F. C., & Carle, R., (2001). By-products of plant food processing as a source of functional compounds-recent developments. *Trends in Food Science and Technology, 12*, 401–413.
57. Sethi, A. P. S., (2012). Effect of tomato pomace on the performance of broilers. In: Dhillon, W. S., Dhatt, A. S., Gill, P. P. S., & Singh, N. P., (eds.), *Proceedings National Seminar on New Frontiers and Future Challenges in Horticultural Crops* (Vol. 1, pp. 210–219). Ludhiana, Punjab Agricultural University.
58. Silva, V. D. M., & Silvestre, M. P. C., (2003). Functional properties of bovine blood plasma intended for use as a functional ingredient in human food. *LWT—Food Science and Technology, 36(*5), 709–718.
59. Simpson, B. K., (2000). Digestive proteinases from marine animals. In: Haard, N. F., & Simpson, B. K., (eds.), *Seafood Enzymes. Utilization and Influence on Postharvest Seafood Quality* (Vol. 1, pp. 191–214). New York, NY, Marcel Dekker.
60. Simpson, B. K., (2007). Pigments from by-products of seafood processing. In: Shahidi, F., (ed.), *Maximizing the Value of Marine* By-Products (Vol. 1(16), pp. 1–27). UK, Woodhead Publishing.
61. Soares, M., Christen, P., Pandey, A., & Soccol, C. R., (2000). Fruity flavor production by *Ceratocystis fimbriata* grown on coffee husk in solid state fermentation. *Process Biochemistry, 35*, 857–861.
62. Toze, S., (2006). Reuse of effluent water. Benefits and risks. *Agricultural Water Management, 80*, 147–159.
63. Turrini, E., Ferruzzi, L., & Fimognari, C., (2016). Potential effects of pomegranate polyphenols in cancer prevention and therapy. *Oxidative Medicine and Cellular Longevity 2015*, 75–83.

64. Varadharajan, V., Vadivel, S. S., & Ramaswamy, A., et al. (2016). Modeling and verification of process parameters for the production of tannase by *Aspergillus oryzae* under submerged fermentation using agro-wastes. *Biotechnology and Applied Biochemistry, 64*, 100–109.
65. Ventura, M. R., Pieltin, M. C., & Castanon, J. I. R., (2009). Evaluation of tomato crop by-products as feed for goats. *Animal Feed Science and Technology, 154*, 271–275.
66. Wadhwa, M., & Bakshi, M. P. S., (2013). *Utilization of Fruit and Vegetable Wastes as Livestock Feed and as Substrates for Generation of Other Value-Added Products* (Vol. 1, pp. 1–33). FAO RAP Publication 2013/04. http://www.fao.org/3/i3273e/i3273e00.htm
67. Wardlaw, D., (2016). Sciatica caused by disc herniation, why is chymopapain chemonucleolysis denied to our patients? *International Journal of Spine Surgery, 10*, 40–44. https://www.ncbi.nlm.nih.gov/pmc/articles/PMC5374990/
68. Yan, H., & Kerr, W. L., (2013). Total phenolics content, anthocyanins, and dietary fiber content of apple pomace powders produced by vacuum-belt drying. *Journal of the Science of Food and Agriculture, 93*, 1499–1504.
69. Zhang, H., & Schroder, J., (2014). Animal manure production and utilization in the US. In, *Applied Manure and Nutrient Chemistry for Sustainable Agriculture and Environment* (Vol. 1, pp. 1–33). Dordrecht, Springer.

# 2

# *Characterization and Treatments of Agricultural Wastes*

## CONTENTS

## 2.1. INTRODUCTION

Agricultural waste is a waste formed as an outcome of numerous agricultural activities. It contains different wastes from farms and manure, slaughterhouses, and poultry houses; fertilizer runoff from fields; pesticides that go into the water; harvest waste, silt, and salt drained from fields; and soils or air. As agricultural wastes may give rise to important environmental difficulties, they have become a growing matter in preceding years; nonetheless, they may also be utilized for various useful functions, as feedstock for energy yield, and for chemical or dye adsorption and chemical recovery. This survey is categorized into four parts given below: waste recycling and reuse, waste minimization and management, waste treatment, and agricultural waste characterization (Figure 2.1).

**Figure 2.1:** Visual representation of different types of agricultural wastes.

*Source: https://www.slideshare.net/natthushrirame/agricultural-waste-management-80609107.*

## 2.2. WASTE CHARACTERIZATION

### 2.2.1. Energy Content

Huang et al. (2011) discovered that waste must have a great H: C ratio and a comparatively less O: C ratio in a thermal enactment of biomass feedstock. It is showed that fruit residuals have greater energy content by a review supervised by Nagle (2011). Results have shown ash contents of 2.2–7.1 wt% db and great volatile-matter content of 67–73 wt.% db, a mean waste-moisture content of 35–75% wb, with heating rates obtained to range between 18.3 to 19.3 MJ/kg. Faraco (2011) found that the residues from the olive tree, cereal crops, grape, and tomato can be regarded as potential raw material for large-scale ethanol extraction because they offered plentiful lignocellulosic waste.

### 2.2.2. Biomass Combustion By-Products

Faraco (2011) analyzed the chemical structure of char produced by burning of agricultural waste (wheat straw and corn straw) and the gasification reactivity, and discovered that the char particles produced under high pyrolysis temperatures which had a greater surface area, many tiny pores with narrower walls and that they undergo some melting. The outcomes have shown that the chars' microcrystals became bigger at higher temperatures. Strong metals, like Hg, can be discharged from biomass combustion. Huang et al. (2011) predicted average annual strong metal discharges from biomass combustion in China (forest and grassland fires, including crop residual burning, etc.) of 27 Mg/year.

### 2.2.3. Agricultural and Natural Fibers

Cocozza et al. (2011) examined chemical, physical, and spectroscopic attributes of the residues of an endemic marine plant, *Posidonia oceanica*. They achieved that due to the lower heating rates between 13.6 and 15.7 Mg/kg, the fibrous portion of *P. oceanica* can be utilized as a biofuel. Additionally, it was discovered it is not appropriate for energy recovery that the leaves likened a green waste with lower heating rates around 10 MJ/ kg (Figure 2.2).

**Figure 2.2:** Visual characteristics of jute fiber.

*Source: https://textilelearner.blogspot. com/2012/03/features-oa-jute-fiber-properties-of.html*

Kuan et al. (2011) estimated the soluble fibrous residues and physicochemical characteristics of residues generated from alkali-treated raw cereal materials (wheat straw and corncob) with great dietary-fiber content (49.87–68.65%). Insoluble fibrous residues and cereal materials were established to carry necessary minerals (in comparison to soluble fibrous residues), and revealed 2,2-diphenyl-1-picrylhydrazyl radical scavenging ability, water capability, oil holding capability, and mineral binding capability, better emulsification ability and emulsion stability. The extraction of nano fibers from SF by electro spinning proved productive. Smith and Jones (2011) discovered that fresh coir fiber has an average consumption coefficient of 0.8 at frequencies higher than 1360 Hz and 20 mm thickness. Raising the thickness can facilitate to enhance lower-frequency sound consumption.

### 2.2.4. Bacterial and Fungal Communities

Zhang et al. (2011) told in detail how fungi and bacteria play a significant role in deciding the formation of agricultural waste. Experimental outcomes have shown that ammonium, nitrate, and WSC affected the temporal fluctuation of bacteria community composition, while fluctuations in fungal community composition was more relying on moisture content, WSC, and pile temperature. Smith and Jones (2011) discovered microbiological impurities originating from the digested product of agricultural by-products and anaerobic co-digestion of bovine manure, in analyses containing the microbiological study of bovine manure, liquid, and solid fractions, saved a liquid fraction of digested product and fresh digested product. The digested product was usually discovered to be extra hygienic than bovine manure. The study also resulted that, this does not rise human health risk in utilizing animal manure for anaerobic co-digestion of bovine manure, agricultural by-products in a field-scale biogas plant and agricultural fertilization.

### 2.2.5. Decomposition

Barrena (2011) investigated through the dynamic respirometric index and oxygen uptake rate the potential for the composition of each sample. Large organic content concentration is unnecessarily instantly linked with degradability. For instance, materials with large lignin concentrations prefer to have less biodegradability. Rather, the respirometric characterization (respirometric index and oxygen uptake rate) could be an importantly more dependable parameter. Arslan (2011) determined the consequence of aeration rates on agricultural waste fertilizing. In the survey, vegetables, and fruit wastes were fertilized at various parameters and rates like as pH, temperature, moisture, electronic conductivity, cellulose, and C/N were checked. He resulted that, the optimum aeration rate for aerobic fertilizing of vegetable-fruit wastes was 0.62 L/min kg VS because the C/N ratio is a significant indicator of stabilization in fertilizing.

In Quebec, research led by Gassara et al. (2011) resulted that the release of greenhouse gases to the atmosphere is caused by the solid wastes from fruits. Hence, whereas current management strategies comprising incineration, composting, solid-state fermentation and landfill generated large value animal feed and enzymes, the generation of greenhouse gases must also be viewed. Particularly, treating 16209 t apple pomace with landfill method concluded in greenhouse gas discharges of 1841 t $CO_2$ eq. per year.

### 2.2.6. Agricultural Water Pollution

Li (2011) discovered that a slow and steady deterioration in the water efficiency of the largest fresh-water lake in China, Poyang Lake with the rise in $COD_{Cr}$, TSS, $BOD_5$, TP, and TN is caused by long-period visibility to agricultural wastewater. A second study projected a figure amounting to 40% of the overall fertilizer N application to China's cropland, the overall waste N emitted to surface water in China at 11.4 kg N in 2008.

Gassara et al. (2011) showed interest in manure used as agricultural fertilizer, which can bring out synthetic and natural hormones to both aquatic ecosystems and agricultural fields. In the Midwestern United States, it is studied that during storm events and effluent irrigation hormone concentrations in subsurface tile flows away were found to raise. Difficulties also emerged to preserve over winter, with pollutant-extractions to sewerage ditches after snowmelt raising hormone concentrations in the water (87 ng/L for total estrogens and 52 ng/L for natural androgens).

## 2.3. REUSE AND RECYCLE

### 2.3.1. Chemical Adsorption

Chen et al. (2011) examined that the highest Zn (II) and Cu (II) adsorption capabilities of corn straw-based biochar attained 12.52 mg/g and 11.0 mg/g, correspondingly, signaling that plant residue-derived biochars can behave as really efficacious sorbents. Acheampong et al. (2011) showed the biosorption of Cu (II) onto sawdust, coconut husk, *Moring oleifera* seeds, and coconut shell. Their conclusions signaled that substitution of K (I) and Mg (II) for Cu (II), electrostatic forces and precipitation were required in the sorption procedure (Figure 2.3).

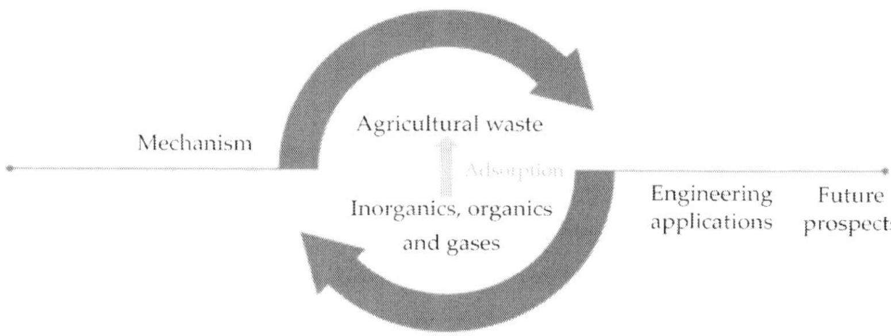


**Figure 2.3:** Lifecycle of agricultural wastes.

Morcali and Aktas (2011) contrasted the gold uptake from water down gold chloride solutions with 100 ppm gold ($Au^{3+}$) utilizing initiated rice husk and Lewatit TP 214. The conclusions revealed initiation energies of 28.44 and 26.41 kJ/ml, and uptake capabilities of 93.3 and 109.6 mg Au/g for initiated rice husk and Lewatit TP 214, correspondingly.

Al-Othman (2011) studied that Cd (II) can be dispatched from aqueous solutions by esterification with maleic acid, utilizing Tamrix articulate wastes altered chemically. The highest adsorption capability of 195.5 mg/g was attained in an adsorbent dose of 0.3 g/L, pH 4 solution at 30°C within the time of 120 min and an initial concentration of 400 mg/L. The sorption procedure was exothermic, constrained, and was greatly represented by a pseudo-second-order equation. Al-Othman et al. (2011) showed the biosorption of Pb (II) from aqueous solution in a batch system utilizing olive tree-pruning waste at pH 5. Constrained and exothermic, the procedure was greatly represented with the pseudo-second-order kinetic model, however, equilibrium conclusions were greatly described with the Sips and Langmuir isotherms, with highest sorption capabilities of 32.15, 33.39 and 26.24 mg/g at 60, 40 and 25°C, correspondingly. Kong et al. (2011) examined the preparation of biochar from soybean stalk utilizing oxygen-limited pyrolysis was efficacious at eliminating Hg (II) and phenanthrene from aqueous solution. Sorption isotherms were greatly represented by Toth model and linear isotherms for Hg (II) and phenanthrene, correspondingly, and when both Hg (II) and phenanthrene were present in the aqueous solution straight competitive sorption was noticed. Talebi and Moussavi (2011) contrasted the co-adsorption of cyanide and Cr (VI) from electroplating wastewater utilizing powdered activated carbon (PAC) and powdered pistachio hull (PHP). With 60 min of contact and 2 g/L of PHP, PHP could totally eliminate cyanide (maximum adsorption capacity of 117.6 and 151 mg/g) and Cr (VI). In comparison, under the same circumstances (highest adsorption capability of 47.6 and 39.4 mg/g), PAC could only eliminate 77.8 and 69.2% of cyanide and Cr (VI), correspondingly.

## 2.3.2. Dye Adsorption

Tepe and Dursun (2011) discovered that an agricultural low-price by-product, dehydrated beet pulp carbon, can be utilized as an efficacious adsorbent to eliminate Chemazol Reactive Red 195 from aqueous solution. The highest dye adsorption capability was received as 58.0 mg/g at 50°C at an optimized pH rate (1.0). Adsorption was discovered to be heavy spontaneous, pH-dependent, and endothermic.

Kumar et al. (2011) discovered methylene blue dye can be consumed onto cashew nutshell from aqueous solution. The external mass shift was the rate-determining step and the procedure was restricted by both pore diffusion and surface. The writers resulted that for the elimination of the methylene blue dye, cashew nutshell is efficacious. Tepe and Dursun (2011) discovered that through adsorption to pistachio hull waste cationic dyes (e.g., methylene blue) can be eliminated from aqueous agricultural waste solutions. The elimination efficacious of methylene blue achieved 99.7% at 70 min contact time when PHP dosage and pH were 1.5 and 8 g/L, correspondingly. Adsorption efficacious raised with raising the temperature. The review resulted that pistachio hull waste is a promising substitution for dye elimination from industrial wastewaters, and hence could be a really efficacious and low-price adsorbent.

Gao et al. (2011) found that Reactive Brilliant Red K-2G can be eliminated by chestnut shells in aqueous solution, a low-price adsorbent. Constrained, adsorption, and exothermic, led a pseudo-second-order procedure with the adsorption value controlled by external mass shift. Hydroxyl, phenolic, and amine compounds were discovered to be energetic binding sites. Parvarthi et al. (2011) revealed that malachite green can be eliminated by cassava peel-based activated carbon from aqueous solution. Outcomes of the initial dye concentration, adsorbent dosage, contact time and pH were viewed, and conclusions signaled that silver nitrate-processed cassava peel conducted smoother than Sulphur acid processed and raw carbons. Gao et al. (2011)

signaled that Ever direct Orange-3GL and Direct Blue-67 dyes can absorb instantly onto rice husk from aqueous solution. The sorption procedure was related to hydroxyl functional-group action and was less affected by either the initial dye amount or the biosorbent dose than by pH.

### 2.3.3. Energy

Chang and Cheng (2011) utilized bagasse for hydrogen manufacturing. They were utilized to hydrolyze natural agricultural waste (i.e., bagasse) for $H_2$ manufacture by utilizing distinct hydrolysis and fermentation procedure because xylanase/cellulases manufactured by *Clostridium pasteurianum* can hydrolyze cellulosic materials efficaciously. The hydrogen generates in this review was approximately 0.96 mol $H_2$/mol glucose.

Gautam et al. (2011) utilized the MixAlco™ procedure to transfer biomass to carboxylate salts biologically which can then be chemically transformed into a broad kind of fuels and chemicals. Agricultural residues or wastes with a large lignocellulose substance (such as bagasse and paper) can be utilized as feedstock and generally have a large generated value of biological fuel as contrasted to less lignocellulose substance agricultural waste. Gautam et al. (2011) contrasted gasification of various biomass feedstock from agricultural waste. After gasification, although poultry litter gasification gave the least heating rate of 4.8 $MJ/m^3$, peanut hull pellets were discovered to have the maximum heating rate of 6.1 $MJ/m^3$,

Keskin et al. (2011) studied the application of photosynthetic bacteria to collect solar energy and to drive the almost complete transformation of substrates to carbon dioxide and hydrogen.

This procedure is also termed as photo-fermentation which can potentially utilize to generate the required value of hydrogen from different agricultural and industrial waste streams abundant in organic acids or sugars as feedstocks.

Malatji et al. (2011) signaled that for electricity production, briquettes made from agricultural and wood waste can be gasified in a downdraft gasifier.

Their conclusions have shown a gasification performance corresponding to solid pine wood. Meeuwse (2011) examined that from agricultural wastes, the potential application of lipid-accumulation fungi for the generation of biodiesel precursors. An example was formulated to report the lipid accumulation, growth, and substrate ingestion of the oleaginous fungus *Umbelopsis isabellina* in submersed chemostat traditions.

Nonetheless, the access may be invalid for lesser dilution values due to practical difficulties with the filamentous fungus. Malatji et al. (2011) examined by utilizing a laboratory compactor for pellet generation, the densification traits of Brazilian rice husks. They discovered that rice husk pallets were stable and long-lasting with 17% water substance and ground with 2, 4, and 6 mm sieves.

Through an energy life-cycle assessment utilizing rice straw as a bio-fuel source, Shie et al. (2011) examined four potential gasification methodologies and various energy predictors. Conclusions signaled that individual gasification methodologies have a favorable energy gain; moreover, major constituents of the energy life cycle analysis were the estimation of the energy inputs for pre-treatment and transportation.

To determine methane generations and reactor behavior, Jose-Cuetos et al. (2011) examined three energy crop residues (sunflower residues or rapeseed, maize) and the co-digestion of swine manure.

Batch practical tests were managed with swine manure having 75, 50 and 25% volatile solids. Conclusions signaled that with the maximum generation attained when co-digesting rapeseed (3.5 L/day), the additional crop residues raised the rate of biogas manufactured.

Omar et al. (2011) examined that they have chemical and fuel traits corresponding to other biomass kinds and described oil palm vacant fruit bunches for microwave-assisted pyrolysis.

High volatile content, moderate potassium content and high-water content were found. The conclusion revealed that for biomass with traits given above corresponding to palm oil vacant fruit bunches, microwaves were the good heating channel.

Tye et al. (2011) demonstrated that second-generation bioethanol as a consumable energy channel, in specific in the transportation region. They forecasted that from agricultural waste, the biomass obtainable is $50.9\times10^6$ kg/year in Malaysia, with the potential to produce $13.3\times10^6$ kg/year of bioenergy-around 21.5% of total Malaysia's energy demand.

Through enzyme hydrolysis and acid, Jose-Cuetos et al. (2011) found the transformation into banana plant waste, corn cobs, and glucose of rice straw for ethanol manufacturing.

By using the steam explosion technique and microwave treatment and chemical pulping, the items were first pre-dealt and then forwarded to acid hydrolysis applying 5% sulfuric acid or enzyme hydrolysis utilizing

*Trichoderma reesei* celluloses. The conclusions revealed the kind of pre-dealt and the hydrolysis technique changed accordingly to the glucose amount and resulting ethanol manufacturing yields. Gu et al. (2011) illustrated the biofuel manufacture utilizing waste nitrogen on marginal or untilled land in China.

In 2008, they predicted that the overall potential biofuel manufactured by waste nitrogen and the overall waste nitrogen wasted to surface water were $16.4 \times 10^{12}$ kJ/year and $11.3 \times 10^{9}$ kg/year, correspondingly.

Presently, the final energy equilibrium of biofuel manufacture utilizing waste N in China was $5.7 \times 10^{5}$ kJ/km$^2$/year-much greater than other main biofuel manufacture systems in use.

In China, the overall highest land demand for the overall energy usage was reviewed to be 17.5% of China's total untilled land asset.

From agri-food waste, Kahiluoto et al. (2011) found that in two areas in Finland by-product biomass and energy rebuilding and nutrient renewing could reduce climate change and eutrophication.

They reported that energy recovery responded to 5–10% of the fossil energy utilized in the area, and followed to a one-third decrease of all agriculture-based GHG discharges.

The decrease in fertilizer production with efficacious renewing attributed to the addition of one-fifth decrease in GHG discharges along with a 14–20% decrease in energy application. Nutrient renewing responded to 45–72% of N and 99% of P in the yields cultivated.

### 2.3.4. Biofertilizers, Cultivation Materials, and Soil Amendments

Graciolli and Figueiro (2011) applied bean, maize cobs, sorghum straw and wheat, banana leaf, and rice as the substrate in nourishing an edible mushroom, *Pleurotus florida*. They discovered that substrates with great levels of K, Ca, P, Mn, and Mg and great ash content, a C/N ratio approximately 45 created the finest nourishment medium. Kahiluoto et al. (2011) examined nitrate leaching from the use of cattle slurry (CS), oat double-cropping process, sewage sludge and urban waste fertilizer to maize.

Various processes involving organic debris with a great C/N ratio, a nitrification inhibitor with a mineral fertilizer process (MIN) and CS were viewed (Figure 2.4).

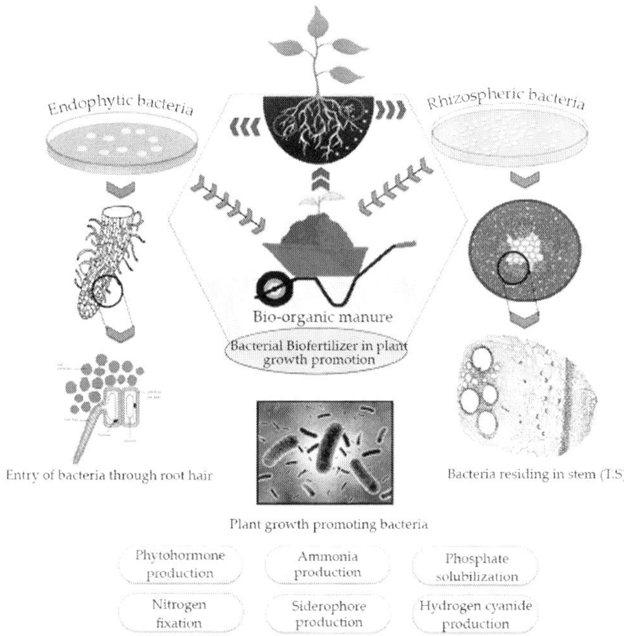


**Figure 2.4:** Overview of mechanism for bio-fertilizers.

*Source: https://www.sciencedirect.com/topics/agricultural-and-biological-sciences/biofertilizers.*

Split uses of sewage slime, (pulp mill sludge—PMS) collectively with mineral fertilizer (PMS+M) and urban waste fertilizer. Split applications of sewage slime, MIN, and PMS+M revealed greater yearly nitrate wastes (145, 145, 152 kg N/ha, correspondingly) contrasted to the operation (50 kg N/ha).

Jacinthe et al. (2011) demonstrated that manure use to the nourished area not only participated to soil carbon concatenation but also raised the nutrients in the soil. They recommended that organic farming exercises raised the scale of the soil microbial community, in contrast with conventional farming exercises, and organic farming procedure added carbon substrate more efficaciously as contrasted with the community acquaint in conventional exercise.

Jacinthe (2011) signaled that vinery soil revised with grape marc vermicompost could be utilized in nitrogen and carbon mineralization. They resulted in that to maintain initial levels of organic matter, 21 kg/km$^2$/year and 17 kg /km$^2$/year of bulge vermicompost should add up if a 2°C increase in temperature and the present refer temperature are viewed, correspondingly.

Kunhikrishnan et al. (2011) found the photo availability and conversion of Cu in Adelaide and Gawler Hills (Australia) soils processed with revived water and Cu usage by Indian mustard (*Brassica juncea var. Varuna*) plants. Pore water-free $Cu^{2+}$ amounts and checking of $CaCl_2$ extractible signaled that raising the level of Cu followed to raised concentrations in plants. Nevertheless, due to the involvement of organic carbon in the renewed water, there is a decrease in the Cu usage contrasted to Milli-Q™ water. On P recovery from animal waste, Rittmann et al. (2011) determined, the demanded transformation of organic P into inorganic P and can be completed through anaerobic treatment.

Zanon et al. (2011) showed to manage bacterial wilt provoked by *Ralstonia solanacearum* by the application of tomato crop residues as a soil revision. Practical tests were implemented with tomato development substrate revised with tomato debris and unnaturally vaccinated with the pathogen. Heat treatments at 45°C reduced tomato wilt indicators and eliminated the pathogen; in comparison, still, after six weeks of constant treatment, fewer temperature treatments at 25°C did not eliminate the pathogen.

Carbonell et al. (2011) demonstrated the impact of municipal solid waste fertilizer (50 Mg/ha) and NPK (nitrogen, phosphorous, potassium) compost (33 g/plant) changes on the number of metals in soils and various portions of maize plants. Municipal solid waste fertilizer and NPK compost increased biomass manufacture, but the solid waste fertilizer enhanced the concentration of Pb, Zn, and Cu in the soil, on the other hand, NPK fertilizer raised Ni and Cd, and reduced the Hg rate. The concentration of metals was greater in roots than in upper-ground portions of the plant, with no important conflicts between the two revisions.

Rosenqvist and Dimitriou (2011) examined that the flagrant margin for brief rotation coppice is merely positive if biomass manufacture is higher than 1.442 t GJ/km$^2$, but the application of wastewater and sewage slime as fertilizer raises the flagrant earning margin by 39 and 199 €/GJ. They predicted that if all slime and wastewater could be utilized to huge plantations of brief rotation coppice in Europe 6000 PJ of revived energy could be manufactured yearly, whereas the application of further P-rich residues might be more affordable.Odlare et al. (2011) investigated the potential risks and benefits of using compost and biogas residues from source-separated household waste in agriculture as compared to traditional mineral fertilizer. They found that biogas residues produced crop yields almost as high as those obtained using a mineral fertilizer. Biogas residues and compost improved substrate-induced respiration, potential ammonium oxidation, and nitrogen mineralization with no apparent negative effects.

In Nepal, Etter et al. (2011) studied the composition of struvite ($MgNH_4PO_4 \cdot 6H_2O$) utilizing urine as fertilizer, alike means for optimizing the procedure. They examined that a reactor with an outward filtration system had a greater P-removal capability (90%), contrasted to a reactor depend on sedimentation only (50%).

Odlare et al. (2011) demonstrated the possible risks and gain of applying fertilizer and biogas residues from source-divided household waste in agriculture in comparison to a cultural mineral compost. They discovered that biogas residues manufactured crop generates nearly as great as those attained utilizing a mineral compost. Fertilizer and biogas residues enhanced substrate-evoked potential ammonium oxidation, nitrogen mineralization and respiration with no visible negative impacts.

With an overall N-content of 0 to 1200 kg N/ha, on the microbial involvement and composition of bacteria in agricultural soil, Nakatani et al. (2011) studied the impacts of tannery slime uses. The primary slime use was examined to arouse biological involvement heavy up to 300 days after use through raises in soil pH and inorganic-N; enzymatic involvements concerned with N-cycling (*urease* and *asparaginase*) were the most affected properties. The second use had a lesser and shorter impact on microbial involvement, guiding to unmeasurable conflict between treatments after 666 days.

On the accessibility of Pb, Zn, and Cu, Medhioub, and Rigane (2011) demonstrated the impact of two municipal solid waste fertilizers (at 3% and 6% dry weight) as amendments for polluted agricultural soil. The conclusions revealed that the fertilizers decreased the leach ability of these three metals and decreased the solubility of the metals, specifically Pb and Cu. Medhioub and Rigane (2011) examined the physicochemical properties of six organic soil additions (sesame bark, mixtures of compost-manure, olive mill wastewater sludge, olive cake, poultry manure, and almond shell) yielded from agricultural waste. The use of these treatments as constituents in reorganized anthropic soils revealed conflicts in pH, electrical conductivity, and soil organic matter content, with subsequent impacts on the efficiency of tomatoes and generation.

## 2.3.5. Chemical Generation and Recovery

From wheat bran hydrolysates and rice straw, Gopinath et al. (2011) studied that amino acid can be manufactured by infusion pentose-utilizing *Corynebacterium glutamicum*. By the wild kind-derived infusion the ethambutol-triggered generation of l-glutamate was approximately up to 93 +/– 4 mM, whereas l-lysine by the infusion lysine generation was approximately up to 42 +/– 2 mM. The conclusion showed that acid hydrolysates of agricultural waste materials by offering carbon and energy sources may serve a replacement feedstock for large-scale amino acid generation

Ohura et al. (2011) determined that orange waste, a cheaper and usable agricultural waste, could be applied in P growth and recovery. Infusion with Zr, the suitability of its use was interrogated for P recovery from side-stream liquid and secondary drainage, which carried 68.2 and 5.9 mg/dm$^3$ P, correspondingly. An active adsorption capability of 1.3 mol P/kg was obtained, after secondary drainage treatment with Zr-SOW gel. Moreover, by utilizing 0.2 M NaOH, the adsorbed P was effectively extracted as a concentrate.

By utilizing corn cob powder as a substrate, Velmurugan et al. (2011) demonstrated that *Monascus purpureus* KACC 42430 can manufacture pigments in solid-state fermentation. They reached to a pigment generate of 25.42 OD Units/g using 60% (w/w) prior moisture content, inoculation with 4 ml of spores/gram of moisture-free substrate, an incubation time of 7 days, and incubation at 30°C. Pigment concentration was greater than those manufactured from several agricultural waste substrates and was steady at high temperatures, in salt solutions, and acidic pH.

From microbial protein of lignocelluloses waste, Khanifar et al. (2011) demonstrated that tryptophan can be manufactured. *Pleurotus florida* was nourished applying solid-state fermentation and wheat straw was processed with the heat at 100°C under alkaline circumstances. Tryptophan was examined utilizing alkaline hydrolyzes with Ba $(OH)_2$, 4 Normal at 110°C for 48 hours of the collected protein. As noticed by an A-200 amino nova tester, tryptophan was 0.96 g/100 g of the collected protein. The study discovered the nutritional gain that could make this individual cell protein a suitable animal feed.

Bajwa et al. (2011) reported the characteristics and appropriateness of residual plant fibers as a strengthener in thermoplastic composites. Five fiber filler processes were interrogated: CBS with 2% (by weight) second cut linters (CBL), control (oak wood fiber filler-OWF), cotton burr and sticks (CBS), CBS with 30% (by weight) guayule bagasse (CGB), and CBS with 30% (by weight) guayule entire plant (GCP). The conclusions revealed that nonetheless, every exhibited potential for application as fiber fillers, the mechanical and physical characteristics of the composites diverged.

Tuan et al. (2011) demonstrated the formulation and characteristics of silver nanoparticles (4–30 nm) use in the process of microbial infection and their burdened in coconut husk-based activated carbon. The nanoparticle-loaded activated carbon had a surface region of 890 $m^2/g$ and revealed alike methylene adsorption and morphology capability contrasted to non-loaded activated carbon, with heavy As (V) adsorption and additional benefits of a less minimal-inhibitory concentration of *Escherichia coli*.

Sarangi et al. (2011) examined the impact of carbon composite powder manufactured from pyrolyzing rice-husk ash and temperature on nano-crystalline silica. The crystallite sizes of carbon and silica and stage transition, as an operation of heat process at temperatures between 400 and 1200°C over two hours, were studied through X-ray diffractometry, scanning electron microscopy and Fourier converts infra-red spectroscopy.

Kim (2011) reviewed techniques to recover salts from rejected brine from desalination plants, agricultural drainage and other industries, including hybrid systems that combine two or more techniques. He also conducted an economic analysis comparing, using, and disposing of the concentrate.

By utilizing a dissolution-precipitation method, Yang, and Liou (2011) studied Nano silica can be manufactured from rice husk. They signaled that hydrochloric acid manufactured silica with a greater surface region than oxalic, citric, and sulfuric acids. Conclusions revealed the greatest surface region of 634 $m^2/g$, pore diameters from 3 to 9 nm, consistent particle scale of 5–30 nm, and the greatest pore volume of 0.811 $cm^3/g$.

Min et al. (2011) contrasted the properties of pectins made from apple pomace applying water-based extraction merged with enzymatic and physical processes and chemically-derived pectins. Pectins manufactured by enzymatic/physical processes had lesser galacturonic acid and generate, a greater degree of esterification, reduced viscosity and low elastic characteristics under great shear-value circumstances than those manufactured through chemical production. However, when exposed to stable shear, pectins derived by both techniques revealed shear-thinning characteristics.

### 2.3.6. Construction Material

Khorami et al. (2011) reported that contrasted bagasse, eucalyptus fiber and wheat to generate fiber cement boards. The conclusions revealed that the flexural attitude of fiber cement boards based on the kind, diameter, length, texture, and aspect ratio of the fibers. It also concluded that raising fiber content from 2% to 4% of the cement weight can raise flexural potency.

Karademir et al. (2011) signaled that yarn waste and luffa fibers could be applied in fusion with renewed cellulose fibers for packaging materials and insulation. The sound absorption, thermal stabilities, glow of the control sheets, print density, tone values, and air permeability were all enhanced; nonetheless, yarn waste and luffa fibers also importantly decreased the tensile indexes and burst (Figure 2.5).

**Figure 2.5**: Construction material prepared from agricultural wastes.

### 2.3.7. Enzyme and Catalyst

By microbial action utilizing thermophilic actinomycetes, Khan et al. (2011) interrogated the manufacture of amylase from agricultural and industrial waste. They calculated that the strain AH-II was the finest for the manufacture of amylases by studying the six strains of actinomycetes (AH-I to AH-VI). The impact of metal ions and optimal circumstances for the highest amylase manufacture were also evaluated.

The potential of the evoked enzyme to saccharify three lignocellulosic substrates and the induction of cellulose enzyme by *Aspergillus oryzae*

ITCC-4857.01, was interrogated by Alimon and Begum (2011). After the alkali process for 96 hours, sugarcane bagasse enhanced specific substrate usage (g dry biomass /0.99 g substrate), decreased sugar generate (7.53 mg/ml) and saccharification value (9.03%), and demonstrated the highest degree of transformation of the substrate (0.415%). Water hyacinth pre-processed with enzyme and alkali revealed the highest endoglucanase action of 11.42 U/ml, meanwhile, sawdust revealed the minimum saccharification value (5.92%) when processed with alkaline and the minimum decreased sugar generate (1.12 mg/ml) when processed with the enzyme.

Chan (2011) revealed that as a catalyst in the formulation of phenols from aryl iodides and CsOH applying CuO on mesoporous silica, rice husks can be utilized. Moreover, the packing of the catalyst was discovered to be as less as 1 mol-%, which creates rice husk really appealing green catalyst.

## 2.4. TREATMENT

### 2.4.1. Anaerobic Treatment

Alimon and Begum (2011) discovered less microbiological and chemical dangers from the an aerobically digested residue (ADR) manufactured in three Norwegian full-scale biogas plants (BGP) and intended for agricultural application.

The ADR revealed great amounts of Cd in the liquid fraction and great amounts of Cu and Zn in the solid fraction. Referring to Norway standards categorizes it as a superior 1 organic fertilizer. Just two out of eleven pesticides sensed, thiabendazol (<0.14–0.73 mg/kg) and imazalil (<0.30–5.77 mg/kg), were discovered commonly, while two contaminants, PAH 16 and diethylhexylphthalate, were discovered in huge amounts.

Madsen et al. (2011) surveyed monitoring techniques for anaerobic digestion treatments. Specifically, they concentrated on the potential use of Theory of Sampling (TOS) and chemometrics, Process Analytical Technologies (PAT), and highlighted the industrial organic waste and biomass, Danish co-digestion of agricultural manure as a scenario.

### 2.4.2. Thermal Treatment

Roy et al. (2011) influenced that metal discharges and gaseous from incineration were lesser than presenting regulations or could be visited with existing methods, making incineration encouraging methods by studying present biosolids management choices. In addition, they studied productive conclusions in a conventional wood pellet stove from pilot-scale burning tests of co-incineration of biosolids (10%) with wood pellets (90%) with the absence of important degradation of efficacious and discharges.

Madsen et al. (2011) signaled that through a ceramization treatment, conversion of dehydrated sewage slime pellets into a lightweight clay material has potential applications in the agriculture industries and building.

The concluding material revealed a microstructure with less thermal conductivity, undetectable levels of risky metals, and open porosity. Contrasted to the generation of conventional clay ceramics the treatment generated similar discharges factors, but greater organic discharges that would demand an air-depuration system.

### 2.4.3. Adsorption Treatment

Through low-price and conventional sorbents, Chen et al. (2011) studied the pesticide-removal theory, including biosorbents attained from agricultural by-products.

They focused on the impact of contact time, Ph, initial pesticide concentration, optimization of biosorption circumstances, and sorbent dosage. Chen et al. (2011) examined the potential of three magnetic biochars, MOP400, MOP700, and MOP250 to eliminate organic contaminants and phosphate at the same point from water. MOP400, MOP700, and MOP250 were made by the subsequent establishment of iron oxide magnetite through pyrolyzation at 250, 400 and 700°C and chemical co-precipitation of $Fe^{2+}$/$Fe^{3+}$ on orange peel powder. MOP400 showed the greatest sorption capacity for organic contaminants, meanwhile, the magnetic biochars demonstrated much lower sorption capacity for non-magnetic biochars than phosphates.

### 2.4.4. Composting

To provide nutrients to *Setaria spendida* L. *grass*, Sabrina et al. (2011) contrasted the impacts of P-enriched vacant oil palm fruit bunches utilized as fresh, vermi fertilized or fertilized media. Soil processed with vermi fertilized and fertilized media concluded in a higher root volume contrasted to the operation, higher positive impact on growth and nutrient consumption, meanwhile soil processed with vermi fertilized vacant oil palm fruit bunches manufactured importantly greater raises in grass moisture-free matter generate contrasted to other soils because of quickly obtainable P and different nutrients (Figure 2.6).

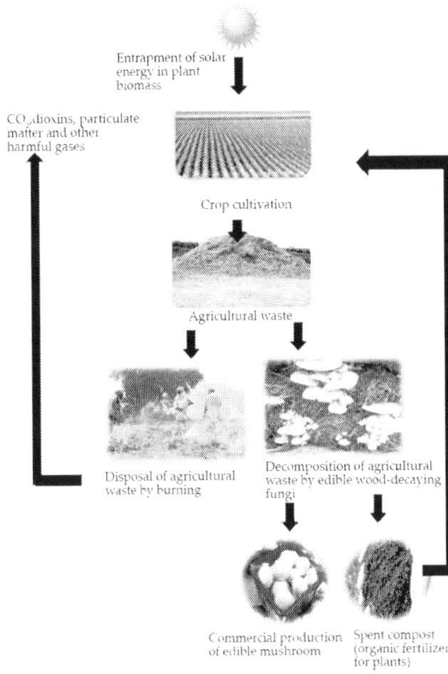


**Figure 2.6:** Composting mechanism by wood decomposing fungi species.

*Source: https://www.researchgate.net/figure/Composting-process-by-wood-decaying-fungi_fig1_316076418.*

The conclusions also revealed that the overall K amount in the control (242.0 mg/kg) was greater than in the vermi fertilized-and fertilized- vacant oil palm fruit bunches (167 and 173 mg/kg, correspondingly). Gallo et al. (2011) examined chicken manure and duck manure, agricultural wastes and kitchen wastes as N amendments in bench-scale fertilizing of renewed paper mill slime. Overall NPK in feedstock changed with manure pleased the needs, but pathogen amounts surpassed the United States Environmental Protection Agency levels. Manure-amended feedstock handled nicer than feedstock amended with agricultural wastes and kitchen wastes at a C/N of 30. Agricultural waste-amended feedstock revealed importantly reduced available air space in the prior levels of fertilizing, meanwhile, Kitchen waste-amended feedstock displayed great ammonia discharges.

Between the period of instant fertilizing of dairy manure with rice chaff, Liu et al. (2011) examined the variations in microbiological and biochemical parameters. The instant fertilizing technique was discovered to be possible for processing agricultural wastes, whereas a dairy manure/rice chaff ratio

(V/V) at 75/25 generated nearly rapid temperature rise, the greatest microbial population, and enzymatic actions, and therefore the greatest fertilizing rate. The C/N ratio was noticed to be less than 20% on day 28 of the practical test.

### 2.4.5. Bioreactor

Stephan et al. (2011) studied the debasement of chlortoluron and metolachlor in a falling film, closed-loop photoreactor as a stage in eliminating pesticides from agricultural wastewater. They pointed out the requirement to enhance the explanation of the mineralization and also studied comforting conclusions in explaining the degradation kinetics.

### 2.4.6. Wetland Treatment

In mesoscale integrated built wetlands over a 13-month time period, Si et al. (2011) interrogated microbial elimination from the liquid fraction of an aerobically digested pig manure by applying four processes: high nutrient loading (T1), high flow rate (T2), effluent recycling (T3), and standard (T4). T1 and T2 demonstrated greater mean counts of molds and yeasts and spore-forming bacteria than T3 and T4. Counts were additionally impacted by flow through the cells and season.

From the Yellow River in Northwest China Si et al. (2011) determined that built wetlands can efficaciously decrease strong metal amounts in irrigation water. They discussed that built wetlands could reduce the amounts of strong metals by up to approximately 35% and that the usage of wheat grain irrigated with unprocessed water presents a potential health danger to the locals.

### 2.4.7. Other Technologies and Assessments

Akbari et al. (2011) investigated an analysis of air and fuel combining in a hydrogen and methane burning system and examined its execution with agricultural waste - or biomass-evaluated fuels. Their computational liquid dynamics structures involved stable turbulence (forms of RANS, Reynolds Average Naiver-Stokes), unstable-RANS, and huge eddy structures. They resulted that they invaded and scattered in a similar manner for a specified momentum flux ratio, although methane and hydrogen have various diffusivity and densities.

An environmental life-wheel evaluation of seven process choices for biogas digestate was reported by Müller and Rehl (2011). They resulted that solar drying, fertilizing, and physical-chemical treatment utilized fewer resources and had a lesser environmental effect than conventional management, from the research of the primary energy need, acidification potential, and global warming potential. N discharges, the value and kind of fuel utilized as heat supply, and the allocation of power and heat were known as significant factors impacting the environmental effect of the management choices.

Ozonated granular activated carbon as an ammonia adsorbent in aqueous systems was determined by Love et al. (2011). The conclusions indicated that from aqueous systems activated carbon could be chemically altered to improve the adsorption of ammonia. Although, the involvement of evaporable organic compounds such as acetic acid and p-cresol suppressed adsorption of the ammonia.

## 2.5. WASTE MANAGEMENT AND MINIMIZATION

Gassara et al. (2011) announced that, of all the present GHG management techniques (landfill, incineration, the solid-state fermentation, and compositing), solid-state fermentation to manufacture enzymes was a nearly efficacious technique for decreasing discharges, while the landfill technique doubled GHG discharges (Figure 1.7).

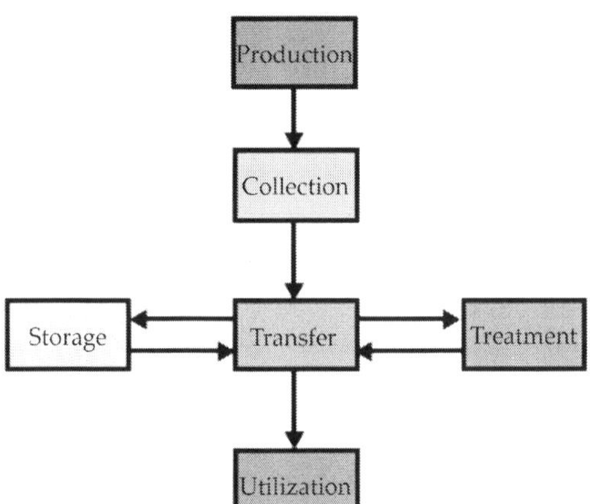


**Figure 2.7:** Agricultural waste management and utilization mechanism.

*Source: https://www.researchgate.net/figure/Agricultural-Waste-Management-Functions_fig1_308880744.*

Chiew et al. (2011) planned an analytical tool that integrates price, $CO_2$ discharges, and energy savings to influence optimal biomass energy manufacture, and used it to examine five possibilities for the palm oil factory in Malaysia. They resulted that the application of extra heat for biomass pre-process, and also the implementation of new Power plants and Combined Heat were advantageous.

In the Philippines Hara et al. (2011) determined present organic waste recycling, and through urban agriculture, he also studied the potential for local renewing. Conclusions signaled that attaining purchasers for fertilizer is hard currently and that nearly fertilize is distressed to huge neighboring farm villages; nonetheless, the formation of a local fertilizer renewing system may be possible.

A biodegradable drip irrigation mechanism, utilizing an irrigation pipe came from a grade of Mater-Bi™, and embedded cylindrical drippers came from an individual Mater-Bi grade was manufactured by Hara et al. (2011). From the irrigation system, it offered a feasible way to release non-biodegradable waste.

In Luxembourg the concentration of dissolved pharmaceuticals and herbicides in a small river was studied by Meyer et al. (2011). The concentration of diuron at 7.41 parts per million in a flood event was calculated by them, which was above 4 times the permissible concentration. In catchments, lower than 50 $km^2$ further severe troubles could originate. Moreover, the measurements evidenced that the impact of kinematic wave results on the relationship between chemograph and hydrograph should not be ignored in smaller basins.

## REFERENCES

1. Acheampong, M. A., Pereira, J. P. C., Meulepas, R. J. W., & Lens, P. N. L., (2011). Biosorption of Cu(II) onto agricultural materials from tropical regions. *J. Chem. Technol. Biotechnol., 86(*9), 1184–1194.
2. Akbari, A., Hill, S., McDonell, V., & Samuelsen, S., (2011). Experimental and computational analyses of methane and hydrogen mixing in a model premixer. *Proc. ASME Turbo Expo: Power Land, Sea, Air*, *133*(10), 843–853.
3. Aktas, S., & Morcali, M. H., (2011). Gold uptake from dilute chloride solutions by a Lewatit TP 214 and activated rice husk. *Int. J. Miner. Process.*, *101*(1–4), 63–70.
4. Al Othman, Z. A., Hashem, A., & Habila, M. A., (2011). Kinetic, equilibrium and thermodynamic studies of cadmium (II) adsorption by modified agricultural wastes. *Molecules, 16(*12), 10443–10456.
5. Arslan, E. I., Unlu, A., & Topal, M., (2011). Determination of the effect of aeration rate on composting of vegetable-fruit wastes. *Clean: Soil, Air, Water, 39(*11), 1014–1021.
6. Bajwa, S. G., Bajwa, D. S., Holt, G., Coffelt, T., & Nakayama, F., (2011). Properties of thermoplastic composites with cotton and guayule biomass residues as fiber fillers. *Ind. Crops Prod., 33*(3), 747–755.
7. Barrena, R., Turet, J., Busquets, A., Farres, M., Font, X., & Sanchez, A., (2011). Respirometric screening of several types of manure and mixtures intended for composting. *Bioresour. Technol., 102(*2), 1367–1377.
8. Begum, M. F., & Alimon, A. R., (2011). Bioconversion and saccharification of some lignocellulosic wastes by Aspergillus oryzae ITCC-4857.01 for fermentable sugar production, *Electron. J. Biotechnol., 14(*5), 1–33.
9. Bermudez, G. M. A., Jasan, R., Pla, R., & Luisa, P. M., (2011). Heavy metal and trace element concentrations in wheat grains: Assessment of potential non-carcinogenic health hazard through their consumption. *J. Hazard. Mater.*, *193*, 264–271.
10. Bouraoui, F., & Grizzetti, B., (2011). Long term change of nutrient concentrations of rivers discharging in European seas. *Sci. Total Environ., 409(*23), 4899–4916.
11. Carbonell, G., Imperial, R. M. D., Torrijos, M., Delgado, M., & Rodriguez, J. A., (2011). Effects of municipal solid waste compost and mineral fertilizer amendments on soil properties and heavy metals

distribution in maize plants (*Zea mays* L.). *Chemosphere, 85(*10), 1614–1623.

12. Chan, C. C., Chen, Y. W., Su, C. S., Lin, H. P., & Lee, C. F., (2011). Green catalysts derived from agricultural and industrial waste products: The preparation of phenols from CsOH and aryl iodides using CuO on mesoporous silica. *Eur. J. Org. Chem.*, (36), 7288–7293.
13. Chen, B., Chen, Z., & Lv, S., (2011). A novel magnetic biochar efficiently sorbs organic pollutants and phosphate. *Bioresour. Technol., 102(*2), 716–723.
14. Chen, X., Chen, G., Chen, L., Chen, Y., Lehmann, J., McBride, M. B., & Hay, A. G., (2011). Adsorption of copper and zinc by biochars produced from pyrolysis of hardwood and corn straw in aqueous solution. *Bioresour. Technol., 102*(19), 8877–8884.
15. Cheng, C. L., & Chang, J. S., (2011). Hydrolysis of lignocellulosic feedstock by novel celluloses originating from pseudomonas sp CL3 for fermentative hydrogen production. *Bioresour. Technol., 102(*18), 8628–8634.
16. Chiew, Y. L., Iwata, T., & Shimada, S., (2011). System analysis for effective use of palm oil waste as energy resources. *Biomass Bioenergy, 35(*7), 2925–2935.
17. Cocozza, C., Parente, A., Zaccone, C., Mininni, C., Santamaria, P., & Miano, T., (2011). Chemical, physical and spectroscopic characterization of *Posidonia oceanica* (L.) Del. residues and their possible recycle. *Biomass Bioenergy, 35(*2), 799–807.
18. Dimitriou, I., & Rosenqvist, H., (2011). Sewage sludge and wastewater fertilization of short rotation Coppice (SRC) for increased bioenergy production-biological and economic potential. *Biomass Bioenergy, 35(*2), 835–842.
19. Dursun, A. Y., & Tepe, O., (2011). Removal of chemazol reactive red 195 from aqueous solution by dehydrated beet pulp carbon. *J. Hazard. Mater.*, *194*, 303–311.
20. Etter, B., Tilley, E., Khadka, R., & Udert, K. M., (2011). Low-cost struvite production using source-separated urine in Nepal. *Water Res., 45(*2), 852–862.
21. Faraco, V., & Hadar, Y., (2011). The potential of lignocellulosic ethanol production in the Mediterranean basin. *Renewable Sustainable Energy Rev., 15(*1), 252–266.

22. Figueiro, G. G., & Graciolli, L. A., (2011). Influence of the chemical composition of the substrate in the cultivation of *Pleurotus florida*. *Cienc. Agrotecnol., 35*(5), 924–930.
23. Gallo, J., Naraval, J. Jr., & Gallardo, S., (2011). Bench-scale composting of recycled paper mill sludge using chicken manure plus duck manure, kitchen wastes and agricultural wastes as nitrogen amendments. *Philipp. Agric. Sci., 94(*3), 292–300.
24. Gao, J. F., Wang, J. H., Yuan, Q., Yang, C., Wang, S. Y., & Peng, Y. Z., (2011). Utilization of agricultural waste chestnut shell for the removal of reactive brilliant red K-2G from aqueous solution. *Desalin. Water Treat., 36*(1–3), 141–151.
25. Gassara, F., Brar, S. K., Pelletier, F., Verma, M., Godbout, S., & Tyagi, R. D., (2011). Pomace waste management scenarios in Quebec-Impact on greenhouse gas emissions. *J. Hazard. Mater., 192*(3), 1178–1185.
26. Gautam, G., Adhikari, S., Brodbeck, C., Bhavnani, S., Fasina, O., & Taylor, S., (2011). Gasification of wood chips, agricultural residues, and waste in a commercial downdraft gasifier. *Trans. ASABE, 54(*5), 1801–1807.
27. Gopinath, V., Meiswinkel, T. M., Wendisch, V. F., & Nampoothiri, K. M., (2011). Amino acid production from rice straw and wheat bran hydrolysates by recombinant pentose-utilizing *Corynebacterium glutamicum*. *Appl. Microbiol. Biotechnol., 92*(5), 985–996.
28. Greene, S., Taylor, D., McElarney, Y. R., Foy, R. H., & Jordan, P., (2011). An evaluation of catchment-scale phosphorus mitigation using load apportionment modeling. *Sci. Total Environ., 409(*11), 2211–2221.
29. Hara, Y., Furutani, T., Murakami, A., Palijon, A. M., & Yokohari, M., (2011). Current organic waste recycling and the potential for local recycling through urban agriculture in metro Manila. *Waste Manage. Res., 29(*11), 1213–1221.
30. Hosseini, F. M., Ayub, M., & Jailani, M. N. M. (2011). Analysis of coir fiber acoustical characteristics. *Appl. Acoust., 72(*1), 35–42.
31. Huang, X., Li, M., Friedli, H. R., Song, Y., Chang, D., & Zhu, L., (2011). Mercury emissions from biomass burning in China. *Environ. Sci. Technol., 45(*21), 9442–9448.
32. Iglinski, B., Iglinska, A., Kujawski, W., Buczkowski, R., & Cichosz, M., (2011). Bioenergy in Poland. *Renewable Sustainable Energy Rev., 15*(6), 2999–3007.

33. Jacinthe, P. A., Shukla, M. K., & Ikemura, Y., (2011). Carbon pools and soil biochemical properties in manure-based organic farming systems of semi-arid New Mexico. *Soil Use Manage., 27(*4), 453–463.
34. Jose, C. M., Fernandez, C., Gomez, X., & Moran, A., (2011). Anaerobic co-digestion of swine manure with energy crop residues. *Biotechnol. Bioprocess Eng., 16(*5), 1044–1052.
35. Kahiluoto, H., Kuisma, M., Havukainen, J., Luoranen, M., Karttunen, P., Lehtonen, E., & Horttanainen, M., (2011). Potential of agri food wastes in mitigation of climate change and eutrophication—Two case regions. *Biomass Bioenergy, 35(*5), 1983–1994.
36. Karademir, A., Aydemir, C., & Yenidogan, S., (2011). Sound absorption and print density properties of recycled sheets made from waste paper and agricultural plant fibers. *Afr. J. Agric. Res.*, 6(28), 6073–6081.
37. Keskin, T., Abo-Hashesh, M., & Hallenbeck, P. C., (2011). Photo fermentative hydrogen production from wastes. *Bioresour. Technol., 102(*18), 8557–8568.
38. Khan, M. A., Shah, M., Qurashi, M. Z., Tanzeel urn, R., & Riaz, S., (2011). Production of amylase from agro-industrial wastes by using thermophilic actinomycetes. *Asian J. Chem., 23(*11), 4947–4954.
39. Khanifar, J., Ghoorchian, H., Ahmadi, A. R., & Hajihosaini, R., (2011). Determination of the amount of tryptophan from single cell protein (SCP) of the lignocelluloses wastes. *Afr. J. Agric. Res., 6(*24), 5434–5437.
40. Khorami, M., & Ganjian, E., (2011). Comparing flexural behavior of fiber-cement composites reinforced bagasse: Wheat and eucalyptus. *Construction and Building Materials, 25(*9), 3661–3667.
41. Kim, D. H., (2011). A review of desalting process techniques and economic analysis of the recovery of salts from retentates. *Desalination, 270(*1–3), 1–8.
42. Kong, H., He, J., Gao, Y., Wu, H., & Zhu, X., (2011). Cosorption of phenanthrene and mercury(II) from aqueous solution by soybean stalk-based biochar. *J. Agric. Food Chem., 59(*22), 12116–12123.
43. Kuan, C. Y., Yuen, K. H., Bhat, R., & Liong, M. T., (2011). Physicochemical characterization of alkali treated fractions from corncob and wheat straw and the production of nano fibers. *Food Res. Int., 44(*9), 2822–2829.
44. Kumar, P. S., Abhinaya, R. V., Lashmi, K. G., Arthi, V., Pavithra, R., Sathyaselvabala, V., Kirupha, S. D., & Sivanesan, S., (2011). Adsorption

of methylene blue dye from aqueous solution by agricultural waste: Equilibrium, thermodynamics, kinetics, mechanism and process design. *Colloid J., 73(*5), 651–661.

45. Kunhikrishnan, A., Bolan, N. S., & Naidu, R., (2011). Phytoavailability of copper in the presence of recycled water sources. *Plant Soil, 348(*1–2), 425–438.
46. Lancelot, C., Thieu, V., Polard, A., Garnier, J., Billen, G., Hecq, W., & Gypens, N., (2011). Cost assessment and ecological effectiveness of nutrient reduction options for mitigating phaeocystis colony blooms in the Southern North Sea: An integrated modeling approach. *Sci. Total Environ., 409(*11), 2179–2191.
47. Liou, T. H., & Yang, C. C., (2011). Synthesis and surface characteristics of nanosilica produced from alkali-extracted rice husk ash, *Mater. Sci. Eng.: B, 176(*7), 521–529.
48. Liu, D., Zhang, R., Wu, H., Xu, D., Tang, Z., Yu, G., Xu, Z., & Shen, Q., (2011). Changes in biochemical and microbiological parameters during the period of rapid composting of dairy manure with rice chaff. *Bioresour. Technol., 102(*19), 9040–9049.
49. Love, C. D., Kolar, P., Classen, J. J., & Das, L., (2011). Adsorption of ammonia on ozonated activate carbon. *Trans. ASABE, 54(*5), 1931–1940.
50. Madsen, M., Holm-Nielsen, J. B., & Esbensen, K. H., (2011). Monitoring of anaerobic digestion processes: A review perspective. *Renewable Sustainable Energy Rev., 15(*6), 3141–3155.
51. Malatji, P., Mamphweli, N. S., & Meincken, M., (2011). The technical pre-feasibility to use briquettes made from wood and agricultural waste for gasification in a downdraft gasifier for electricity generation. *J. Energy South. Afr., 22(*4), 2–7.
52. Meeuwse, P., Tramper, J., & Rinzema, A., (2011). Modeling lipid accumulation in oleaginous fungi in chemostat cultures: I. Development and validation of a chemostat model for umbelopsis isabellina. *Bioprocess Biosyst. Eng., 34(*8), 939–949.
53. Meyer, B., Pailler, J. Y., Guignard, C., Hoffmann, L., & Krein, A., (2011). Concentrations of dissolved herbicides and pharmaceuticals in a small river in Luxembourg. *Environ. Monit. Assess., 180*(1–4), 127–146.
54. Minghui, L., Wen, Z., Yu, X., & Yongsheng, G., (2011). Study on removal efficiencies of pollutant from constructed wetland in

aquiculture waste water around Poyang Lake. *Procedia Environ. Sci., 10* (Part C) 2444–2448.

55. Mouri, G., Takizawa, S., & Oki, T., (2011). Spatial and temporal variation in nutrient parameters in stream water in a rural-urban catchment, Shikoku, Japan: Effects of land cover and human impact. *J. Environ. Manage., 92(*7), 1837–1848.
56. Moussavi, G., & Talebi, S., (2011b). Comparing the efficacy of a novel waste-based adsorbent with PAC for the simultaneous removal of chromium (VI) and cyanide from electroplating wastewater. *Chem. Eng. Res. Des. 1*, 1–30.
57. Nagle, M., Habasimbi, K., Mahayothee, B., Haewsungcharern, M., Janjai, S., & Muller, J., (2011). Fruit processing residues as an alternative fuel for drying in Northern Thailand. *Fuel, 90*(2), 818–823.
58. Naik, A. P., & Hammerschmidt, C. R., (2011). Mercury and trace metal partitioning and fluxes in suburban Southwest Ohio watersheds. *Water Res., 45(*16), 5151–5160.
59. Nakatani, A. S., Martines, A. M., Nogueira, M. A., Fagotti, D. S. L., Oliveira, A. G., Bini, D., Sousa, J. P., & Cardoso, E. J. B. N., (2011). Changes in the genetic structure of Bacteria and microbial activity in an agricultural soil amended with tannery sludge. *Soil Biol. Biochem., 43(*1), 106–114.
60. Odlare, M., Arthurson, V., Pell, M., Svensson, K., Nehrenheim, E., & Abubaker, J., (2011). Land application of organic waste—Effects on the soil ecosystem. *Appl. Energy, 88(*6), 2210–2218.
61. Ohura, S., Harada, H., Biswas, B. K., Kondo, M., Ishikawa, S., Kawakita, H., Ohto, K., & Inoue, K., (2011). Phosphorus recovery from secondary effluent and side-stream liquid in a sewage treatment plant using zirconium-loaded saponified orange waste. *J. Mater. Cycles Waste Manage., 13*(4), 293–297.
62. Omar, R., Idris, A., Yunus, R., Khalid, K., & Aida, I. M. I., (2011). Characterization of empty fruit bunch for microwave-assisted pyrolysis. *Fuel, 90(*4), 1536–1544.
63. Parvathi, C., Maruthavanan, T., Sivamani, S., & Prakash, C., (2011). Biosorption studies for the removal of malachite green from its aqueous solution by activated carbon prepared from cassava peel. *E-J. Chem., 8*, 61–66.
64. Pereira, B. C. K. B., Megumi, K. M. C., & Dantas, V. M. C., (2011). Antimicrobial activity and mineral composition of shiitake mushrooms

cultivated on agricultural waste. *Braz. Arch. Biol. Technol., 54(*5), 991–1002.

65. Piemonte, V., & Gironi, F., (2011). Land-use change emissions: How green are the bioplastics? *Environ. Prog. Sustainable Energy, 30(*4), 685–691.
66. Rafiee, E., Khodayari, M., Shahebrahimi, S., & Joshaghani, M., (2011). 12-Tungstophosphoric acid supported on nano silica from rice husk ash as an efficient catalyst for direct benzylation of 1,3-dicarbonyl compounds in solvent-free condition. *J. Mol. Catal. A: Chem., 351*, 204–209.
67. Ramesh, S. T., Gandhimathi, R., Badabhagni, N., & Nidheesh, P. V., (2011). Removal of Cd (II) from aqueous solution by adsorption onto coir pith, an agricultural solid waste: Batch experimental study. *Environ. Eng. Manage. J., 10(*11), 1667–1673.
68. Rehl, T., & Muller, J., (2011). Life cycle assessment of biogas dig estate processing technologies. *Resour., Conserv. Recycl.*, *56*(1), 92–104.
69. Rigane, M. K., & Medhioub, K., (2011b).Assessment of properties of Tunisian agricultural waste composts: Application as components in reconstituted anthropic soils and their effects on tomato yield and quality. *Resour., Conserv. Recycl.*, *55*(8), 785–792.
70. Rigane, M. K., Michel, J. C., Medhioub, K., & Morel, P., (2011a). Evaluation of compost maturity, hydrophysical and physicochemical properties: Indicators for use as a component of growing media. *Compost Sci. Util., 19(*4), 226–234.
71. Rittmann, B. E., Mayer, B., Westerhoff, P., & Edwards, M., (2011). Capturing the lost phosphorus. *Chemosphere, 84(*6), 846–853.
72. Roy, M. M., Dutta, A., Corscadden, K., Havard, P., & Dickie, L., (2011). Review of biosolids management options and co-incineration of a bio-solid-derived fuel. *Waste Manage., 31(*11), 2228–2235.
73. Sabrina, D. T., Hanafi, M. M., Mahmud, T. M. M., & Azwady, A. A. N., (2011). Evaluation of nutrients released from phosphorus-enriched empty oil palm fruit bunches as growing media using *Setaria* splendida. *Compost. Sci. Util., 19(*1), 61–68.
74. Sadasivam, S., Kandasamy, K. S., Ponnusamy, K., Nagarajan, G. S., & Kang, T. W., (2011). Film diffusion mechanism of methyl parathion in biopolymers: Kinetics and thermodynamic equilibrium. *J. Chem. Eng. Data, 56(*11), 4024–4030.

75. Saleh, K. A., & Alshehri, M. A., (2011). The intensity of pollutant genotoxicity in Lake Uluabat: Investigation of the peripheral erythrocytes of *Cyprinus carpio*. *Afr. J. Biotechnol., 10(*71), 16045–16050.
76. Sarangi, M., Nayak, P., & Tiwari, T. N., (2011). Effect of temperature on nano-crystalline silica and carbon composites obtained from rice-husk ash. *Composites, Part B. 42(*7), 1994–1998.
77. Schuh, M. C., Casey, F. X. M., Hakk, H., DeSutter, T. M., Richards, K. G., Khan, E., & Oduor, P. G., (2011). Effects of field-manure applications on stratified 17 beta-estradiol concentrations. *J. Hazard. Mater.*, *192*(2), 748–752.
78. Shie, J. L., Chang, C. Y., Chen, C. S., Shaw, D. G., Chen, Y. H., Kuan, W. H., & Ma, H. K., (2011). Energy life cycle assessment of rice straw bio-energy derived from potential gasification technologies. *Bioresour. Technol., 102(*12), 6735–6741.
79. Si, W., Ji, W., Yang, F., Lv, Y., Wang, Y., & Zhang, Y., (2011). The function of constructed wetland in reducing the risk of heavy metals on human health. *Environ. Monit. Assess.*, *181*(1–4), 531–537.
80. Silveira, F. D. M., Favaretto, N., Dieckow, J., Pauletti, V., Vezzani, F. M., & Borges, D. S. E. D., (2011). Liquid dairy manure in on-till: Carbon and nitrogen loss by runoff. *Rev. Bras. Cienc. Solo*, *35*(5), 1759–1767.
81. Stephan, B., Ludovic, L., & Dominique, W., (2011). Modeling of a falling thin film deposited photocatalytic step reactor for water purification: Pesticide treatment. *Chem. Eng. J., 169*(1–3), 216–225.
82. Taylor, M. A., & Fraser, P. D., (2011). Solanesol: Added value from *Solanaceous* waste. *Phytochemistry, 72(*11–12), 1323–1327.
83. Tian, Y., Wu, M., Lin, X., Huang, P., & Huang, Y., (2011). Synthesis of magnetic wheat straw for arsenic adsorptionm *J. Hazard. Mater.*, *193*, 10–16.
84. Tuan, T. Q., Son, N. V., Dung, H. T. K., Luong, N. H., Thuy, B. T., Anh, N. T. V., Hoa, N. D., & Hai, N. H., (2011). Preparation and properties of silver nanoparticles loaded in activated carbon for biological and environmental applications. *J. Hazard. Mater.*, *192*(3), 1321–1329.
85. Tye, Y. Y., Lee, K. T., Abdullah, W. N. W., & Leh, C. P., (2011). Second-generation bioethanol as a sustainable energy source in Malaysia transportation sector: Status, potential and future prospects. *Renewable Sustainable Energy Rev., 15(*9), 4521–4536.

86. Velmurugan, P., Hur, H., Balachandar, V., Kamala-Kannan, S., Lee, K. J., Lee, S. M., Chae, J. C., et al., (2011). Monascus pigment production by solid-state fermentation with corn cob substrate. *J. Biosci. Bioeng.*, *115*(6), 590–594.
87. Wu, Y., He, J., Hu, Z., Yang, L., & Zhang, N., (2011). Removal of UV(254) (nm) matter and nutrients from a photobioreactor-wetland system. *J. Hazard. Mater.*, *194*, 1–6.
88. Xuan, L. I., Baotong, D., & Hua, Y. E., (2011). The research based on the 3-R principle of agro-circular economy model-the Erhai lake basin as an example. *Energy Procedia, 5(*0), 1399–1404.
89. Yates, S. R., Knuteson, J., Zheng, W., & Wang, Q., (2011). Effect of organic material on field-scale emissions of 1,3-dichloropropene. *J. Environ. Qual., 40(*5), 1470–1479.
90. Zanon, M. J., Font, M. I., & Jorda, C., (2011). Use of tomato crop residues into soil for control of bacterial wilt caused by *Ralstonia solanacearum*. *Crop Prot., 30(*9), 1138–1143.
91. Zeng, G., Zhang, J., Chen, Y., Yu, Z., Yu, M., Li, H., Liu, Z., Chen, M., Lu, L., & Hu, C., (2011). Relative contributions of archaea and bacteria to microbial ammonia oxidation differ under different conditions during agricultural waste composting. *Bioresour. Technol., 102(*19), 9026–9032.
92. Zhang, C., Kang, H., Lv, K., Chen, H., & Yuan, S., (2011). Maize phytoliths and photoluminescent silica nanotubes prepared from a natural silica resource. *Ind. Eng. Chem. Res., 50(*22), 12521–12526.
93. Zhang, J., Zeng, G., Chen, Y., Yu, M., Yu, Z., Li, H., Yu, Y., & Huang, H., (2011). Effects of physico-chemical parameters on the bacterial and fungal communities during agricultural waste composting. *Bioresour. Technol., 102(*3), 2950–2956.
94. Zhou, H., Loeffler, D., & Kranert, M., (2011). Model-based predictions of anaerobic digestion of agricultural substrates for biogas production. *Bioresour. Technol., 102(*23), 10819–10828.
95. Zhou, Q., Gong, W., Yang, D., Xie, C., Li, Y., Liu, X., Bai, C., & Wang, R., (2011). Assessment of the biosorption characteristics of a spent cottonseed husk substrate for the decolorization of methylene blue. *Clean: Soil, Air, Water, 39(*12), 1087–1094.

# 3

# *Potential Use of Animal by-Products for Food Systems*

## CONTENTS

## 3.1. INTRODUCTION

Over the previous decade, there has been a huge growth in the usage of animal by-products. Because of its potential, in many countries, to food insecurity and combat protein malnutrition. Inedible or edible parts are separated out of the animal by-products after slaughtering. 55% of the production comprises the edible part whereas the other part is inedible by-products (IEBPs). For agricultural and industrial uses, sustainable products can be produced by these IEBPs. The deficiency of feed materials and current costs have a high race between animals and humans, they can be relieved by the efficient usage of animal by-products (Chikwanha et al., 2017; Byers et al., 2002). Environmental pollution will also be decreased by this in society. Lesser production cost, cheaper feed, and reduction in competition, in this way, can be obtained by properly making use of animal by-products, like rumen digesta (Irshad and Sharma, 2015). Concerning the stability of the products and food security, nevertheless, gaps are appearing that should be more focused. This is why the effectiveness and efficacy of utilizing animal by-products as substitute sources of feed ingredients, and the forces related to their products to enhance the performance of livestock in the industry at large, are focused in this chapter (Sans and Combris, 2015; Liu, 2002).

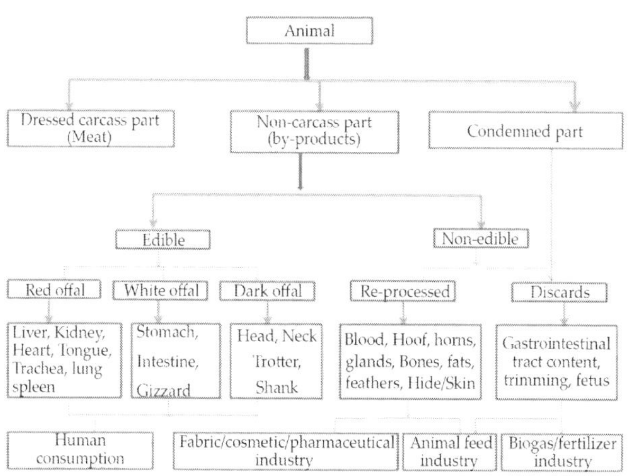


**Figure 3.1:** Categorization of animal by-products.

*Source: https://www.mdpi.com/2071-1050/9/7/1089.*

After testing in the abattoir, EBPs products are advised safe for human utilization as accepted by registered public health inspectors. IEBPs, on the other hand, are used as secondary by-products and are convicted as re-processed or discards, humans cannot eat them. Meat and meat products are restricted to fulfill the human's nutritional requirements; however, important nutrients can be supplied by most EBPs, mainly the kidney, liver, heart, and tongue.

Malnutrition, particularly in the developing countries, is caused by the extensive problem of climate change and food insecurity that the world has to experience today (Fayemi et al., 2016; Sheehy et al., 2005).

Basic food nutrients such as minerals and vitamins have been noted to be lacking in over two billion people in the world, especially in developing countries (including vitamin A, iodine, iron, and zinc).

Around 7.3 kg of animal protein (fish, meat, milk, and egg) per year or 20 g per person per day will be needed to fulfill the nutritional requirements of these people (FAO, 2014).

The danger of food insecurity and malnutrition, however, can be decreased by making use of protein sources like EBPs. The publicity and interpretation of the studies including the use of EBPs as food ingredients have been suggested (Rosa et al., 2002).

Therefore, it has been stated by Ockerman and Basu (2004) that key nutrients like proteins, vitamins (folic acid, B1, B2, and B6), fats, and minerals, with valuable poly-unsaturated fatty and amino acids that are in analogous to those in muscular tissue, are part of EBPs.

Beneficial and suitable products for livestock and human consumption, whereas, can be created by IEBPs like hides and skin, bones, hooves, feathers, hair, horns, bristles, and rumen digesta (Figure 3.2).

It is well acknowledged that leather, upholstery, and textile industry can be based on skin/hide and feathers whereas livestock feed (source of minerals) can be obtained from the bone.

As because of increasing cost of feed, extreme climate condition and competition for cereal crops between livestock and humans, there is a deficiency of feed ingredients in majority developing countries, nonetheless, a substitute feed source for the livestock industry can be presented by the application of rumen digesta (Elfaki et al., 2014).

The significance of animal by-products in the food system is therefore brightened in this review as alternative feed ingredients for the livestock industry and sources nutrient for humans.

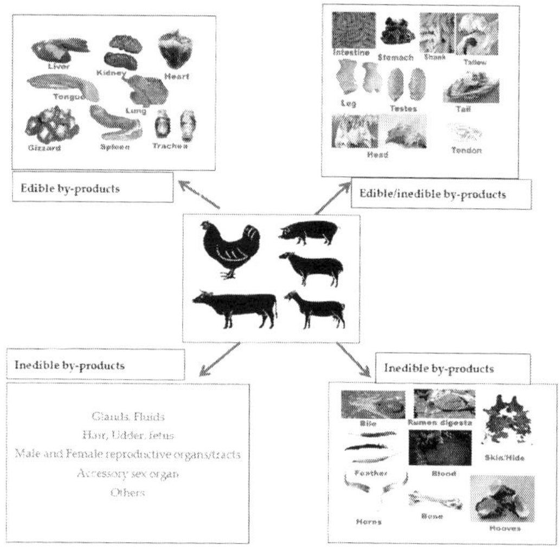


**Figure 3.2:** Generic depiction of inedible and edible by-products.

*Source: https://www.mdpi.com/2071-1050/9/7/1089.*

The animal by-products are categorized and characterized differently according to different usages in the meat industry and vary from land to land. They can be allocated as non-organ and organ by-products or EBPs and IEBPs (Figure 3.1), (Marti et al., 2012). Occasionally, they are characterized as color, shape, and muscular structure. Nevertheless, dark offal, white offal, and red offal are different classes of the offal (Pérez-Alvarez et al., 2011; Alexandre et al., 2010).

## 3.2. TYPES OF DAIRY LEFTOVERS AND NUTRIENT CONTENT

In the grassland system, the nitrogen recovery rate of the dairy slurry is highly inconstant due to changes in slurry composition, spreading rates, application methods, climatic, and soil conditions and slurry nitrogen mineralization rates. In Ireland, nearly 80% of manures produced in the winter season are controlled as slurries comprising 70 g $kg^{-1}$ dry matter (DM), 3.6 g $kg^{-1}$ total nitrogen and 0.6 g $kg^{-1}$ total phosphorus. About 50% of the total nitrogen is in the form of ammonia and can be volatilized as ammonia during the stage storage and subsequent land spreading. Projected organic managed generation of waster for Ireland is given in Table 3.1.

**Table 3.1:** Projected Agricultural Organic Managed Generation of Waste in 2001

| **Waste Category** | **Waste Generation (Tonnes of wet weight)** | **%** |
|---|---|---|
| Water (dairy only) | 18,377,550 | 30.5 |
| Cattle manure and slurry | 36,443,603 | 60.6 |
| Silage effluent | 1,139,231 | 1.9 |
| Pig slurry | 2,431,819 | 4.0 |
| Sheep manure | 1,336,336 | 2.2 |
| Poultry litter | 172,435 | 0.3 |
| Spent mushroom compost | 274,050 | 0.5 |
| Total | 60,170,025 | 100 |

Great discrepancy in the content of nutrient of dairy slurry prevails depending on the type of feed, age of sample when examined, age of the animal and how the waste is stored and managed. Seasonal differences in the contents of the nutrient also exist. Tables of reported slurry contents of nutrients in Europe prevail. In the state of Ireland, dirty water is produced from dairy parlor water and the machine washings, drizzle, and the water from concreted holding patches. The average production of dirty water per cow is around 49 $L^{-1}$ $day^{-1}$. Even though dilute, dirty water has adequate nutrients to give upsurge to eutrophication if gone to the water body through excess infiltration or runoff. Implementation of present legislation needs separation of water and fecal matter, therefore diminishing the content of nutrients of dirty water for the land application. As the content of nutrients is decreased and storage and the water charges are high, a substitute solution to the management of dirty water is remediation and re-utilization for washing yards.

Prediction of the content of nutrients of the agricultural wastewaters will help farmers to precisely calculate the nutrient value of fertilizer replacement of land spread materials and the surplus fertilizer necessities for their crops. Fecal microorganisms

Waste of agriculture not only carry a threat to the water bodies, but a second main concern is also the existence of pathogenic or antibiotic-resistant bacteria in the animal wastes and the danger to human health. If handled and treated properly, manure is an efficient and safe fertilizer. However, if improperly treated or untreated, manure might become the main source of pathogens that might contaminate soil, water bodies and food-stuffs. Animal dungs are known to comprise pathogenic bacteria, parasites, and viruses. The pollution of surface waters with the pathogenic micro-organisms conveyed from fields to which the livestock slurries and dung have been smeared is a severe environmental concern as it might lead to the humans being exposed to microorganisms through bathing waters; drinking water; and water utilized for irrigation of the ready to eat foodstuffs.

Research from New Zealand exhibits that dirty water includes fecal micro-organisms, which instigate from the dairy cattle excreta. The Pathogen Transmission Routes Research Program in New Zealand exhibited that substantial feces pollution arose through the removal of feces by the grazing animals with entrance to waterways. Fencing and execution of buffer strips were suggested as mitigation measures in order to avoid such losses. The existence of fecal pointer organisms is utilized to recognize waters obstructed by fecal matter from the mammals. Indicators of fecal pollution like *E. coli* are broadly utilized as they are supposed to not last for more than four months post excretion. Recent research has displayed that *E. coli* can last for long periods in the temperate soils and assist in high exposures in the drainage water from agricultural soils. *E. coli* were particularly linked with poorly drained soils because of the greater perseverance of superior flow channels and the anaerobic micro-sites where they may survive. Therefore, the existence of *E. coli* in the waters might not indicate current pollution by fecal matter but can be because of the historical pathogen deposition. Several treatment systems might be utilized to treat livestock wastes and decrease or remove viral, eukaryotic, and bacterial pathogens.

## 3.3. CONSUMPTION AND NUTRITIONAL COMPOSITION OF EDIBLE BY-PRODUCTS

It is necessary that prior to the refinement for human use, public health inspector should check the EBPs right after slaughter and should be accepted of not having physical abnormalities and infections (like fibrosis, fasciolosis, tuberculosis, echinococcosis, abscess, and hydatidosis) (Yibar et al., 2015; Gonulalan et al., 2004). Entire organs or just the damaged parts are commonly stated dangerous for human utilization or rejected if any organs and carcasses are infected or anomalies and diseases like wounds are found. To demolish any harmful bacteria that might pose danger to health or infect the consumers, by-products like the stomach and intestine are exhaustively washed then heat-treated, also to get rid of the dirt (Nollet and Toldrá, 2011). Color, surface integrity and imperfections are often used to categorize products like tongue, it includes warts, cuts on the membrane and diseases or physical abnormalities (Figure 3.3).

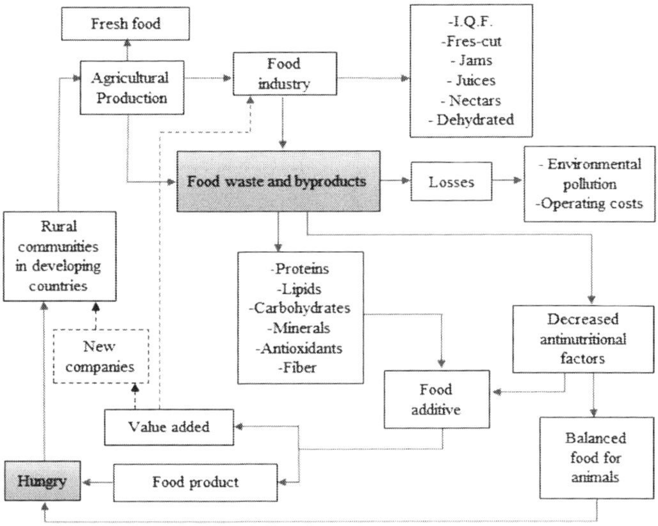


**Figure 3.3:** Flowchart for different stages of production, processing, and utilization of edible by products.

*Source: https://www.frontiersin.org/articles/10.3389/fsufs.2018.00052/full.*

For human advantage, because of finding the importance of EBPs as a nutrients supply, its uses are slowly increasing all across the world. In South America, Europe, Asia, North America, Australia, and Africa, they are broadly used as classic eatables (Toldrá et al., 2012). All pieces of EBPs are treated and widely used as human food in Africa. Jointly between black South Africans and white South Africans (especially Afrikaners), EDP elegance is appreciated and acknowledged as one of the cultural standards. People of all ages traditionally agree and utilize EBPs, as stated by a study conducted in Somalia (Masese and Waweru, 2017; Vanheerden and Morey, 2014). EBPs are commonly grilled, fried, braised with vegetables and other ingredients or boiled before consuming. Many kinds of liver meet have been prepared and used in some areas, for example, foie gras, liver pâté and liver sausages (Xiong et al., 2016). This is for the reason that fat, protein, copper, iron, and vitamins are essential for normal development in people which are found in great concentration in animal liver (Lorenzo and Pateiro, 2013; Abu-Salem and Abou Arab, 2010).

Recently, it was highlighted by 43% of the persons that they eat chicken livers in a survey of 1030 people from the United Kingdom. The research also showed that people of ages from 18 years onwards eat dishes made from chicken liver. On the basis of nutrition, it has been stated that in analogy with muscle tissue, the essential eatable sources of Coenzyme Q10 (vitamin) is beef heart and liver (Ercan and El, 2011; Jones et al., 2016). Moreover, the coenzyme Q10 digestibility in muscle tissue was prominently lower than the coenzyme Q10 digestibility in beef heart and liver. In mitochondria, coenzymes Q10 are produced chiefly, this means that as compared to other parts of muscle tissue mitochondria is contained more by heart and liver muscle.

It is right accepted that compared to other muscular tissue, the liver has more vitamins and minerals and is an essential eatable organ which is 1–2% of the live weight of cattle (Li et al., 2014). It has been observed that at a wet basis, about 3.5% to 7.8% lipid is contained by the beef liver. In general, as compared to the ruminant muscle, less palmitic acid (16:0) and more longer-chain polyunsaturated fatty acids (PUFA), stearic acid (l8:0) is contained by lipids from the ruminant liver (Enser et al., 1998; Chan, 1995). It has also been described that the richness of PUFA in the liver is ten times greater than that of the muscles of the same animals. It was also discovered by Seong et al. (2014) that the muscle tissues of cattle were worthy of comparison to the fat content of pork liver. Fats play an important supervisory role in various biological functions and work as membranes building blocks.

Vital working characteristics, that are connected with physiochemical conditions as ionic strength, pH, and high level of water-soluble protein and with contrasting protein portions, are part of the liver protein (Steen et al., 2016). Nonetheless, it has also been noted that as compared to *longissimus dorsi muscle,* a larger value of tri-unsaturated (C18:3) and di-unsaturated (C18:2) fatty acid is contained by sheep liver. Likewise, saturated (C16:1, C16:0) fatty acid values in the liver were lower than the longissimus dorsimuscle. Purchas and Aungsupakorn (1993) discovered that as compared to beef muscle there is comparatively more iron content in beef liver. Deficiency of iron can result in anemia, especially in children and pregnant women, it is required for accurate movement of blood in the human body (Miller, 2013). It has also been noted that compared to other muscle tissues the liver consists of a balanced quantity of carnosine, taurine, and creatine.

It has been stated that heart muscles have a large amount of leucine, threonine, tryptophan, and lysine and other amino acids. It was given that the amino acid profile and proximate composition of chicken meat are similar to that of gizzard (Arafa, 1977). From poultry, gizzards can be fried and served or eaten in combination with other food ingredients or treated to manufacture fester sausage. Chicken gizzard has less fat (0.9%) and more ash (1.3%) and protein (26%). Sausage casings are usually formed of the bovine intestine (Nakyinsige et al., 2012). It has been recorded that there is a concentrated amount of fat, protein, and mineral content (iron and phosphorus) in viscera organs and blood from goat similarly as in goat muscle tissue. It is also well known that alimentary advantageous characteristic vitamins, minerals, and protein are provided by the tongue (Tremblay, 2011; Madruga et al., 2007;). It has a high concentration of zinc, iron, vitamin B-12 and choline; its consumption revealed that around 12% and 28% of the daily iron uses set for women and men respectively, can be served by cooking each cut of cow tongue which enhances the amount of iron in the body. This gives that before consumption, different treatments like boiling, smoking, curing, seasoned with onion and other spices can be used to treat pork, beef, goat, sheep, and veal tongues. As a nutritional food source in Africa, North America, Asia, and Europe, the cow tongue is used as a primary elegance (Ranken, 2000). On organoleptic and composition characteristics of the cooked cow tongue, still, the study is short.

The comparative abundance of pork and beef genitals in iron, lipid, polyunsaturated acids and protein has been researched. 2.9% fat, 10.5% protein, and 1.3% carbohydrate are found in pork testicles (Pucciarelli et al., 2012).

It has been reported that the presence of the heme iron in pork loin is minimum 2.5 times lower than that in pork testicles (2 mg/100 g, uncooked weight), and it can give 25% of the United States Department of Agriculture (USDA) everyday advised intake for male adults.

To increase male strength, abilities, and manhood, in many parts of the world the penis and testicles from livestock are ordinarily eaten (Rotenberg, 2008).

Spleens are widely used in the treatment of foods in the US and UK, and it is added in processing the sandwich "pani ca meusa," or "bread with spleen" and caciocavallo cheese in Sicily. Nevertheless, because of nutritional value, availability, cultural background, price difference and religious belief, offal is consistently favored to meat.

The requirement for offal is increasing with the rapidly growing population, in regards to the personal approach. In this view, on the uses of EBPs, more studies should be conducted on people's understanding. The complete description will be presented by this on factors affecting offal needs.

## 3.4. FACTORS INFLUENCING THE YIELD AND QUALITY OF EDIBLE AND INEDIBLE ANIMAL BY-PRODUCTS

### 3.4.1. Edible By-Products

Various mutual aspects of genotype (breed), animal species, sex, condition of health and physiology of the animal, production system (feedlot or pasture-based system), weight, age, treatment, and post-mortem handling are the things on which the features and characteristics of eatable animal by-products depend. Florek et al. (2012) reported differences in animal weight at slaughter causes differences in the weight of the by-products, which highlights that as compared to the low body weight more weight of by-products is manufactured by class with high body weight gain and animal of the same breed.

Florek et al. (2012) described that as compared to the tongue, liver, kidney, and heart from baby calves the veal from calves has higher energy, fat, minerals (Ca and Zn), ash, oxidative stability, and fatty acids. On the close structure of all the organs meat, a vital effect of the breed is discovered by Hoffman et al. (2013) in their research, as compared to Dorper sheep, wool sheep have higher fat content in the tongue and liver. As compared to the wool sheep there are higher elements of protein in Dorper heart, brain, spleen, kidney, and testicles. It was also observed that as compared to the liver of merino there are more levels of tyrosine, valine, phenylalanine, isoleucine, and leucine in Dorper liver. According to Li et al. (2014) as compared to Qinchuan cattle, there are more important dissimilarities in the fat, protein, total energy, carbohydrate, amino acids, mineral, glycogen, and PUFA principles of meat of liver from Wagyu × Qinchuan cattle nurtured on the same field.

It was discovered by Puschner et al. (2004), moreover, that with respect to sex, age, and production system the mineral content (copper) of calves is altered where in comparison with beef calves, the quantity of copper in the liver was more in dairy calves. Furthermore, initially, for the first 2 months of age, the richness of copper in the liver for both beef and dairy calves is inclined and then when its amount gradually increases, they are decreased till it approaches around 9–10 months of age. Zawacka et al. (2016) researched that as compared to cockerels, Capons chicken had lesser heart weight and more weight of gizzard and liver. Zawacka et al. (2016) have discovered, moreover, that difference in weight can be controlled by the animal's physiological condition, this result is in analogy with the result concluded by Abdelmageed et al. (2014) that as compared to male chickens, the gizzard weight of female broiler chickens was lighter. As compared to cattle heart, liver, and spleen, in contrast, buffalo had lighter livers and heavier lungs, hearts, spleens, and kidney fat.

### 3.4.2. Effect of Thermal Processing and Preservation Methods on Quality of Edible By-Products

To take command of the deteriorative changes and microbiological growth, it is commonly accepted that EBPs should be stored in refrigerators and are easily rotten. The dirt is removed from the EBPs with running water, then suspended in room temperature for cooling, then it is drip, dried, and chilled at a temperature less than 7°C within 16 h after the slaughter. This is done in order to avoid any decay by microorganisms or degeneration. For 7 days, EBPs can be refrigerated at 0–2°C in many cases, which doesn't have any harmful effect on the product. Liver meat can also be stored for about 4, 12 and 18 months, at freezing temperatures of−12,−18 and−24°C, respectively (Cohen et al., 2006; Meatupdate, 2002). Even though there are some writers that described that there are different microorganisms in EBPs from cattle, sheep, and pig, like *Staphylococcus aureus*, *Escherichia coli*, *Clostridium perfringens*. Still, these preservative mechanisms ensure the safety of consumers of EBPs (Lee and Lee, 2016). Because of their undisciplined storage and treatment, further, this has been known as an unhealthful feature of the meat (Figure 3.4).

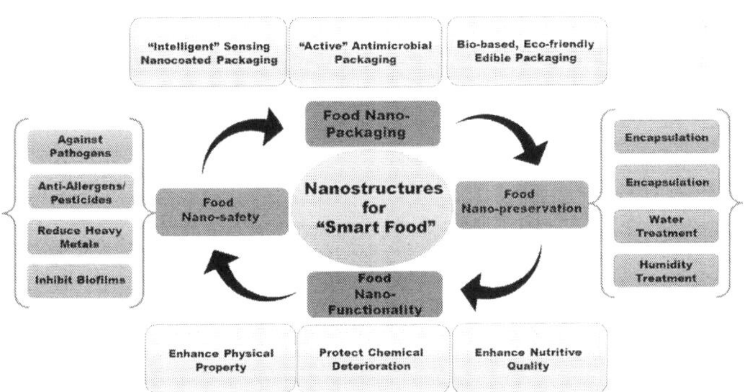


**Figure 3.4:** Modern methods of food preservation using nanotechnology.

*Source:https://www.sciencedirect.com/science/article/pii/S10219498183011 69.*

During preservation in refrigerators, studies are still not completed on the effects of packaging on the quality of eatable offal. To incline the market value and the quality of frozen food, it is needed, however, that the meat is properly packaged, treated, boxed, and frozen for import/export. Culturally, the EBPs are either roasted, boiled, fried, smoked or micro waved before consumption. EBPs are cooked to better the palatability and condition of the meat and to prevent the microbial impurities. Boiling the EBPs is the easiest and commonly used treatment, out of all the processes. On the inside tissues and the exterior surfaces of fresh liver, impurities like Campylobacter have been included especially in the liver of lamb and poultry. To avoid the food-borne infection, proper temperature is needed for cooking, to ensure the safety for consumption. The parasites may not be demolished completely by irregular heating from microwave but still, when the temperature within the product approaches at about 67°C, the destruction of cyst in meat could be done successfully. Recently, the studies on cooked liver showed that mineral structure of the liver from cattle nurtured on-field diets and the fatty acid is not prominently affected by heat at 80 and 60°C (Falowo et al., 2017). Except for this report, on the standard of EBPs, no other scientific studies are conducted on the heat treatment effects.

## 3.5. UTILIZATION OF INEDIBLE BY-PRODUCTS

As discussed before, discards that are depicted by IEBPs are harmful as human food, as shown in Figure 3.1, they are retreated as secondary products (belts, gelatin, and keratin extraction, pharmaceuticals, and footwear) or discarded as wastes (contents of stomach and blood). Common uneatable animal by-products are hair, hides, teeth, horns, fats, blood, ligaments, and cartilage, bone, manure, feet, rumen contents, trimmings, and glands (Figure 3.2). Secondary and elementary by-products are subcategories of IEBPs (Sharma et al., 2013). Elementary by-products involve bones, blood, intestine, pancreas, hoofs, hides, and skins and horns, they are being separated after slaughter (Figure 3.2). Secondary by-products may include bone meal, buttons, cutlery handles, insulin, fat, tennis strips, and gelatin and are derivatives from elementary by-products. Covered from slaughterhouses are often transformed into the leather by using tanning agents for manufacturing secondary by-products such as clothing, shoes, car seats, and other products after being exported to various regions like Asia, South America, Europe, and North America for again treatment (DAFF, 2012). Glands are other IEBPs that are essential to the growth of accessible recent human medicines and of the best standard. It has been worked out that these days many of the enzymes and hormones used in the medical field sustainability depends on organs and glands, they are accumulated from slaughtered animals like parathyroid, adrenal, thyroid, pituitary, pancreas, ovaries, and testes (Aberle et al., 2001). Medicinal products such as estrogens, epinephrine, insulin, progesterone, parathyroid hormone, trypsin, somatotropin, adrenocorticotropic hormone (ACTH), testosterone, thyroid-stimulating hormone, and thyroxin can be produced from animal glands. The extra accruing rate of rejected items can result from the loss of potential incomes due to the under-utilization of these by-products (Wismer-Pedersen, 1988).Blood is like a red fluid that is made up of a mixture of plasma and a cellular fraction (white blood cells, red blood cells, and platelets). About 6–7% of the lean meat content of the dead body and 4% of animal live weight has it. Blood has a concentrated amount of proteins, iron, and bioactive compounds in it, these are appropriate for livestock and human use (Bah et al., 2013; Davidson, 2014). Blood proteins are readily consumable, accessible as frozen, liquid, dried, and other forms. Blood is utilized as color enhancers, binding agents, fat replacers, emulsifiers, and meat curing agents in the food industry. Due to the leavening and foaming features of plasma products (spray-dried plasma), they have been well joined into the food industry (Hsieh and Ofori, 2011; Ofori and Hsieh, 2014). Many cultures intake blood either directly as food or with other ingredients like black puddings, blood sausage, and pancakes. To feed pets and livestock, blood is utilized in the making of blood meals in the feed industry. For the development of the raw material for pharmaceutical porphyrin derivative production, probiotic bacteria (bovine plasma), the active ingredient in topical surgical hemostatic applications, a precursor to thrombin production and purification, and a reagent for routine blood clotting in serology laboratories, blood is usually treated and utilized as a medium in medicine (Bah et al., 2016). When blood from any animal is separated, before it can be consumed, it is mandatory that such animals should be free from any infection. There are some particular groups of people, who prohibit blood-related items for ethical, religious, cultural, and/or health reasons.

## 3.6. INDUSTRIAL USAGE OF IEBPS

### 3.6.1. Biogas Industry

A huge amount of solid and liquid waste products are generated by the slaughterhouse, if they are not properly organized, severe environmental degradation can happen (Arvanitoyannis and Ladas, 2008). It has been discussed that by thermal recycling in power plants, the wastes from the slaughterhouse can be used as secondary fuel. However, slaughterhouses solid waste and meat preparation is consumed as fuel in the cyclonic combustor, for biogas and electric power production, the poultry litter, cattle manure and pig waste are also essential (Virmond et al., 2011).

## 3.6.2. Cosmetic and Fabric Industry

It is important to treat the skin so that enzymatic and microbial decomposition is avoided after the skin and hides are removed from a slaughtered animal in the slaughterhouse. Unprocessed materials for industrial treatment and shipping are provided by the by-products like skin and hide in many countries. Shoes, cosmetics, car seats, belts, keratin, bags, and gelatin are ordinary items from skin and hides (Benjakul et al., 2009) (Figure 3.5).

**Figure 3.5**: Commonly used animal By-Products in cosmetics industry.

*Source: https://www.dailyinfographic.com/makeup-animal-products.*

### 3.6.3. Pharmaceutical Industry

Biochemical and chemical draw outs in animal organs and glands are fascinating and extremely good items in the different industries including food, pharmaceutical, and cosmetics, contained as a share in the combination of hormones, enzymes, vitamins, and pigments. Some basic research reveals that many different medical situations such as building healthy red blood cells, hepatitis, preventing liver damage, treating chronic liver diseases, treating certain cancers, regenerating liver tissue and other viral infections can be treated by utilizing these draw outs (Karthikeyan et al., 2007). Minerals and vitamins like copper, calcium, iron sold and phosphorus are there in the liver excerpt in the shape of tablet or powder. Small intestine and heparin from the lungs are utilized to avoid blood clotting during surgeries and transplants and is a prescribed anticoagulant drug (Table 3.1). The adrenal gland is to manufacture steroid drugs, low sugar or overdose is treated by glucagon where to treat diabetic patients, insulin from the pancreas is used (Hekal, 2014) (Figure 3.6).

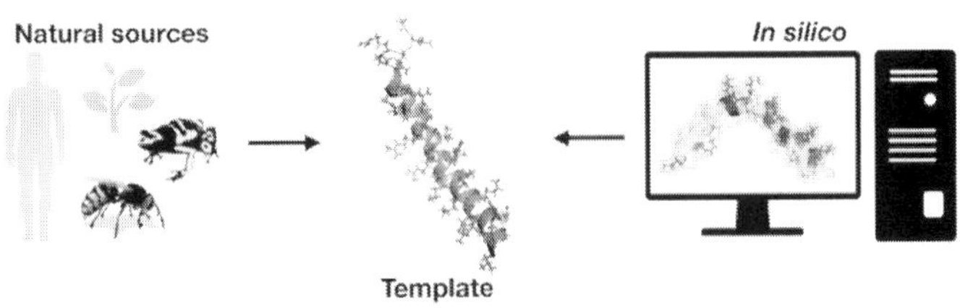


**Figure 3.6**: Reprogrammable biological peptides for treatment of infectious diseases.

*Source: https://pubs.rsc.org/en/content/articlelanding/2019/cc/c9cc07898c.*

**Table 3.2:** Ordinary Applications of Animal **By-Products** (Deivasigamani & Alagappan, 2008)

| **Animal By-Products** | **Hides and Skin** | **Hoof and Horns** | **Bone** | **Blood** | **Intestine** | **Organs & Glands** | **Hair/ Wool** |
|---|---|---|---|---|---|---|---|
| **Reprocessed Products** | Cured hides & skin. Leather & Textiles | Hoof & horn meal Gelatin and keratin extraction | Extraction of collagen Bone meal | Blood meal Pharmaceutical products | Surgical sutures Sausage casings Musical instruments | Medicinal Xenotransplantation Pharmaceuticals | Extraction of keratin textiles |

| | | | | | | | |
|---|---|---|---|---|---|---|---|
| **Major Applications** | Leather clothes, car, and household upholsteries, belts, bags, drums, footwear, luggage, sports goods, wallets, gelatin, etc. | Combs, plates, buttons, souvenirs, Collagen, Fertilizer, glue, foaming in fire extinguishers, gelled food products | Cutlery handles, bone gelatin, Shortening, Collagen, bone meal | Tennis strips, Catgut, pudding or blood sausages, animal feeds, fertilizers, emulsifier, and stabilizer | Musical strings, sports guts, collagen sheets, prosthetic materials, burn dressing, sausage casings, strings for musical instruments, human food, meat meal, pet food, tallow, casings | Heart stimulant, corticotrophins, heparin, steroids, enzymes, estrogen, insulin, progesterone, parathyroid hormone, trypsin | Woven fabrics or cloths, mattress, carpets, keratin, insulators, knitted apparels |

## 3.7. UTILIZATION OF BLOOD AND DRIED RUMEN DIGESTA AS FEED INGREDIENT

Rumen contents also called "digesta," are treated as waste, but they can emerge into a product of high economic value and great importance for the industry of livestock when used or retreated (Liu et al., 2013). To counterpart, the dominant restricted resources of feed, great freedom as a secondary source of nutrients is proposed in a slaughterhouse by the restoration of rumen digesta from ruminant animals. Various researches support the idea that rumen digesta should be added in a specific amount with other ingredients during feed formulation instead of feeding alone to livestock (Adedipe et al., 2005). Nutritionists are more concerned about using them as inexpensive feedstuff due to the chemical composition of blood meal, dried rumen digesta (DRD) and their mixture (Togun et al., 2010). Under other conditions, if blood and rumen content is not completely controlled, they can damage the environment, and result in groundwater pollution, air pollution, and eutrophication. Environmental pollution can be decreased and the flexibility of feed formulation can be increased due to the transformation of these IEBPs into livestock feed (Phua et al., 2016).

On the growth performance of the animals, it is known that there are no disadvantages of making use of DRD as feed ingredients. A reliable source of feed ingredients in catfish ration, poultry ration, lamb ration, quail ration, and cattle ration, is a combination of rumen content and blood which are vital with no unfavorable effects on animals. Environmental pollution is decreased to a minimum with proper treatment and the number of microorganisms in the DRD is also decreased (Patrucco et al., 2016).

## 3.8. NUTRITIONAL BENEFITS OF FEEDING RUMEN CONTENT TO LIVESTOCK

Before using Rumen digesta in feed formulation, they are generally treated by utilizing sun-drying or light heat. The protein content of the ration can be enhanced by blending dry rumen digesta (DRD) with dried blood when utilized as feed constituents. The accessibility of other compounds and amino acid contents are preserved after treating based on the strength of heat used on the rumen digesta (Osman and Elimam, 2015). On different animals, at different stages, dry rumen digesta has been carried out as feedstuff (Table 3.2). In the judgment of DRD North Africa (Sudan and Egypt), Central Africa (Cameroon), West Africa (Nigeria), East Africa (Ethiopia), South East Asia (Thailand), Asia Middle East (Saudi Arabia) and South Asia (India) are the regions where most of the researches have done on the performance of livestock. Due to different chemical composition of preferred fields eaten by different animal species, the DRD may have different proximate composition. Crude protein (CP) of DRD is from 11.38 to 19.6%, dry matter (DM) from 13.36 to 98.4% and crude fiber (CF) 15.3 to 41.84% (Table 3.2). In this regard, increase palatability, value addition and enhance better efficiency is given by mixing DRD with other feedstuffs (Scobie et al., 2015) (Figure 3.7).

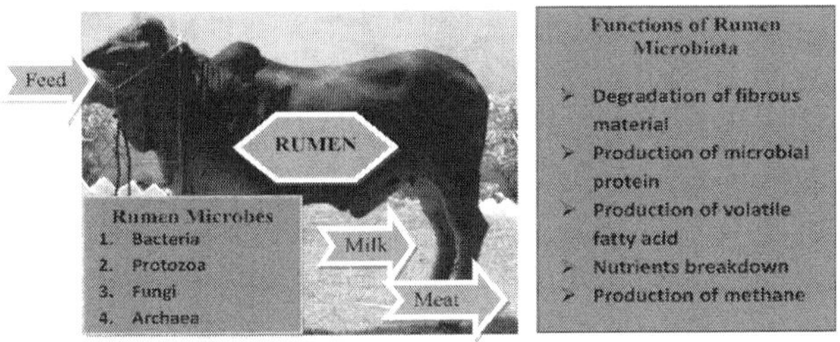


**Figure 3.7:** Estimated ecosystem for rumen microbes.

*Source: https://www.intechopen.com/books/yeast-industrial-applications/probiotic-yeast-mode-of-action-and-its-effects-on-ruminant-nutrition.*

**Table 3.3:** Chemical Composition Belonging to Cattle Dried Rumen Contents Coming Out of Various Countries

| Country | Cameroon | Egypt | India [A] | India [B] | Nigeria [A] | Nigeria [B] | Saudi Arabia |
|---|---|---|---|---|---|---|---|
| Animal Type | Broiler | Duck | Quail | Goat | Rabbit | Broiler | Lamb |
| DM% | 92.44 | 91.3 | 92.64 | 90.5 | 91.8 | 81.8 | 13.36 |
| CP% | 15.21 | 18.53 | 18.26 | 12.8 | 11.38 | 18.52 | 14.2 |
| CF% | 41.84 | 28.28 | 24.99 | - | 24 | 15.3 | - |
| EE% | 7.81 | 7.81 | 3.6 | - | 6.1 | 8.79 | 1.7 |
| ASH% | 9.25 | 9.25 | 14.47 | - | 8.11 | 7.6 | 11.6 |
| NDF% | - | - | - | 78.7 | - | - | 59.2 |
| NFE% | - | 35.97 | - | - | 42.21 | 38.39 | - |
| P% | - | 0.69 | - | - | - | - | - |
| CA% | - | 0.7 | - | - | - | - | - |
| ADF% | - | - | - | -54.5 | - | - | 36.7 |

DM = dry matter, CP = crude protein, CF = crude fiber, EE =ether extract, NDF = neutral detergent fiber, NFE = nitrogen-free extract energy, P = phosphorus, Ca = calcium, ADF = acid detergent fiber.

### 3.8.1. Broilers and Layers

Contrasted to the control group, in the study by Ørskov (2007) showed that broiler birds fed a rumen digesta mixture and a diet consisting dried bovine blood at different levels of 20%, 15%, 10%, 5%, and 0% usually had improved gain in body weight, feed consumption, nutrient utilization values, and organ weight.

Likewise, the broiler chickens fed content blood meal and sun-dried rumen at 15%, 10%, and 5% levels had greater carcass generates and lesser feed cost per unit increase in weight than the function group was noticed by Yitbarek et al. (2016). Because of the rise in crude protein (CP) content and high fiber of the diet, Dairo et al. (2005) illustrated that increase in weight and feed usage of the broilers fed with DRD raised, which made the birds eat to a greater extent just to meet up with the required energy for development and swift growth. Accordingly, in the diets of broiler, the application of DRD as supplements offers an inexpensive price of feed per kilogram of meat manufactured. This concept, which was given by Balogun and Adeniji (2001), is useful as a replacement source of feed in economic means. Compared to the control group, in a study, Jayathilakan et al. (2012)

successfully discovered that a combination of castor oil seed cake with DRD at 25% inclusion and the diet with 5% DRD inclusion level appreciably affect the customer preference and juiciness of broiler meat. As an outcome of this, birds that were fed castor oil seed cake at 15% inclusion in the diet functioned exceptionally good, while, the meat standard signaler of broilers fed with DRD is desirable to that of the broilers in the balanced diet. Hence, without or with enzyme inclusion, the tenderness, color, acceptability, juiciness, and flavor of the carcass standard of the boilers fed with DRD were similar to each other. Moreover, there was no harmful impact on the biochemical and performance characteristics of plasma in broiler chicks, when dry rumen digesta was supplemented in broiler diets with other feed constituents.

The conclusions revealed that birds fed with 15% rumen content with blood meal had lower feed usage than the birds fed with dried rumen content with blood meal at 10%, 5%, and 0% which is somehow related to the study given by Deveau et al. (2004) where the chemical composition of dried rumen content with blood meal and its impact on broiler functions were attempted.

The functions of the broilers because of the given reason showed that the daily increase in weight, ultimate body weight, and feed conversion ratio were higher at a 10% inclusion level. This helps us to reach the conclusions that for all rumen content with blood meal, the feed cost per unit increase in weight was inexpensive, meanwhile, mortality was unaffected.

### 3.8.2. Sheep

Recently, as the inclusion level of DRD in the diet raises, a coherent raise in feed consumption and feed conversion productivity in sheep was noticed by Elimam and Osman (2015). Considering this report, sheep fed had the greatest increase in weight with 10% DRD, meanwhile in the animals fed weight increase was the least with no dried rumen content. There were no important disputes between the weights, and the ultimate body weights were 31.75, 31.25 and 30.27 in animals fed 10%, 5%, and 0% DRD, correspondingly.

### 3.8.3. Cattle

As feed ingredients in cattle ration, the substitution of soya bean meal with 100% DRD (DRD) benefited to strengthen the consumption of rice straw contrasted to other diets was also discovered by Hekal (2014).

### 3.8.4. Lamb

In their study, it was demonstrated by Scobie et al. (2015) that contrasted to the control group, there is no important contrast in the hot carcass weight, dressing percentage, and growth performance of lambs fed with a dietary barley mixture and the mixture of rumen contents at ratio 1:4.

### 3.8.5. Quail

It was revealed by Deveau et al. (2004) that the inclusion of the ruminal digesta and solar dried blood at different levels of 0%, 5%, 10%, 15%, 20%, 25%, and 30% for the substitution of soybean meal in Japanese quail diets. The conclusion showed a greater weight increase as contrasted to the control group.

### 3.8.6. Rabbit

The replacement of maize with DRD at 0%, 12.5%, 25% levels in rabbit diet concluded that there is no important difference in the hematological parameters, and when contrasted with control, the study signaled that the DRD diet was capable to render the needed nutrition to sustain health status of the experimental animals, and the normal blood parameters. Patrucco et al. (2016) examined a corresponding study that the rabbits were fed with a combination of rumen content and bovine blood at various inclusion levels. A raise in daily feed consumption was noticed among the processes but there was no discovery of important disputes in the daily increase of weight of the rabbits.

### 3.8.7. Goat

Togun et al. (2010) examined in a study that at various levels in goat diets, rumen digesta was augmented with wheat offal, and the conclusions showed that due to the aftertaste of the diets all the goats had greater DM consumption. Hence, because of the low inclusion level of DRD, high metabolizable energy, and improved feed conversion ratio the goats that were arranged on the 25% of DRD with 75% wheat offal had the greatest daily increase in weight.Additionally, the study examined the pathogenic potential of rumen contents, and the protection of DRD in animal feed was also evaluated. Mondal et al. (2013) demonstrated that because of the deficiency of the enterotoxin gene, the rumen bacteria are non-pathogenic and that is why if consumed by animals it will not be dangerous. Considering this, when *E. coli* was recorded in post-dumped rumen content, *Staphylococcus aureus* raised in number, and there was an early rise in the number of *E. coli* which reduced afterward, though the microbial load was under the suggested level. Due to this reason, it was resulted by Mondal et al. (2013) that rumen bacteria are non-pathogenic because of the deficiency of the enterotoxin gene, and that is why if consumed by animals it will not be dangerous (Bhaskar et al., 2007).

## 3.9. CONCERNS ASSOCIATED WITH THE USE OF UNEATABLE BY-PRODUCTS

For over forty years, from the renewing of animal waste, the making of animal feed has been in process to lessen the price of feed. Earlier, before the pandemic of Bovine Spongiform Encephalopathy (BSE) in 1988, it was an everyday practice in Europe and other countries to provide the animal carcasses. Bovine Spongiform Encephalopathy titled neurodegenerative disorders are associated with Transmissible Spongiform Encephalopathy (TSE). By a prion agent, the contamination with animal protein dirt through the consumption of feed is the common point of BSE infection. The alteration of normal cellular proteins into a pathogenic kind that spoils the central nervous system of animals is a prion, an infectious protein. Prions are accountable for the obtained kind of neurodegenerative disease and are impermeable to processes that disintegrate nucleic acid (Shin and Hyun, 2000). Generally, in highly sensitive parts of the central nervous system (spinal cord and brain) and bone marrow prions are often discovered.

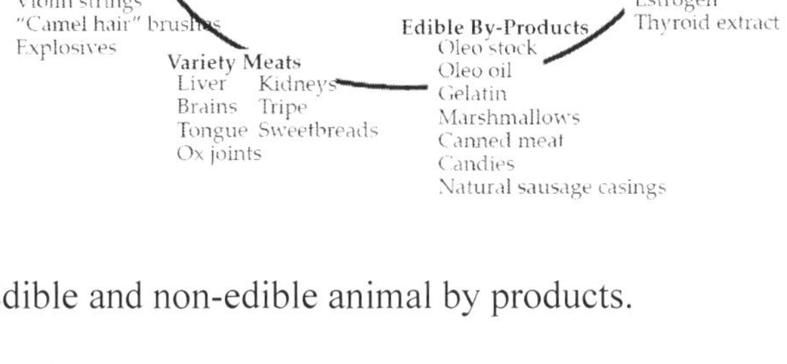


**Figure 3.8:** Edible and non-edible animal by products.

*Source: https://slideplayer.com/slide/11756849/.*

The symptoms of BSE in cattle are: aggression, lack of interaction, change in mood, loss of appetite, and decreased milk yield. In this view, infected animal tissues assist as the prion carriers are identified in many studies. Due to this reason, in order to prevent the disperse of BSE disease to humans and animals usually during manufacturing in the abattoir, particular risk materials, for instance, the nerves attached to the brain, spinal cord, and the brain are eliminated and isolated from meat. Moreover, the application of spinal cords and brains from cattle of any age or cattle bigger than the age of 30 months not passed and inspected for human usage should be banned in the feed or food of all animals is proposed by the Food and Drug Administration (FDA) agency in 2005. This scheme allows the consumption of other uneatable by-products, particularly rumen digesta in animal feed, and has been productive in managing the disperse of BSE. Either unprocessed or processed rumen digesta carries prions that can aid the disperse of BSE and it is yet to be well recognized (QMS, 2010).

## REFERENCES


1. Abdelmageed, M. E. I., Sulieman, A. M. E., Abdalla, H. O., & Salih, G. E., (2014). Effects of incorporating chicken's gizzards and abdominal fat in the quality of burger meat product. *J. Microbiol. Res., 2*, 68–71.
2. Aberle, E. D., Forrest, J. C., Gerrard, D. E., Mills, E. W., Hedrick, H. B., Judge, M. D., & Merkel, R. A., (2001). Conversion of muscle to meat and development of meat quality. In: *Principles of Meat Science* (Vol. 4, pp. 83–107). Kendall and Hunt: Dubuque, IA, USA.
3. Abu-Salem, F. M., & Abou, A. E. A., (2010). Chemical properties, microbiological quality and sensory evaluation of chicken and duck liver paste (Foiegras). *Grasas Y Aceites*, *61*, 126–135. .
4. Adedipe, N. O., Sridhar, M. K. C., Verma, M., & Wagner, A., (2005). Waste management, processing, and detoxification. *Ecosyst. Hum. Well-Being Policy Responses, 3*, 313–334.
5. Adeniji, A. A., & Balogun, O. O., (2001). Evaluation of blood-rumen content mixture in the diets of starter chicks. *Niger. J. Anim. Prod., 28*, 153–157.
6. Alexandre, G., Liméa, L., Nepos, A., Fleury, J., Lallo, C., & Archimede, H., (2010). The offal components and carcass measurements of Creole kids of Guadeloupe under various feeding regimes. *Development*, *22*, 5–33.
7. Arafa, A. S., (1977). Pickled chicken gizzards 1. Acceptability and proximate analysis. *Poult. Sci., 56*, 1014–1017. .
8. Arvanitoyannis, I. S., & Ladas, D., (2008). Meat waste treatment methods and potential uses. *Int. J. Food Sci. Technol., 43*, 543–559. .
9. Bah, C. S., Bekhit, A. E. D. A., Carne, A., & Mcconnell, M. A., (2013). Slaughterhouse blood: An emerging source of bioactive compounds. *Compr. Rev. Food Sci. Food Saf. 12*, 314–331. .
10. Bah, C. S., Bekhit, A. E. D. A., Carne, A., & Mcconnell, M. A., (2016). Composition and biological activities of slaughterhouse blood from red deer, sheep, pig and cattle. *J. Sci. Food Agric., 96*, 79–89. .
11. Benjakul, S., Oungbho, K., Visessanguan, W., Thiansilakul, Y., & Roytrakul, S., (2009). Characteristics of gelatin from the skins of bigeye snapper, *Priacanthus tayenus* and *Priacanthus macracanthus*. *Food Chem., 116*, 445–451. .
12. Bhaskar, N., Modi, V. K., Govindaraju, K., Radha, C., & Lalitha, R. G., (2007). Utilization of meat industry by-products: Protein hydrolysate

from sheep visceral mass. *Bioresour. Technol., 98*, 388–394. .

13. Byers, T., Nestle, M., Mctiernan, A., Doyle, C., Currie-Williams, A., Gansler, T., & Thun, M., (2002). American cancer society guidelines on nutrition and physical activity for cancer prevention: Reducing the risk of cancer with healthy food choices and physical activity. *CA Cancer J. Clin., 52*, 92–119. .
14. Chan, W., (1995). Meat Poultry and Game. In: *Fifth Supplement to the Fifth Edition of McCance and Widdowson's. The Composition of Foods* (Vol.1, pp. 1–22). Royal Society of Chemistry: London, UK.
15. Chikwanha, O. C., Vahmani, P., Muchenje, V., Dugan, M. E., & Mapiye, C., (2017). Nutritional *enhancement of sheep meat fatty acid profile for human health and wellbeing. Food Res. Int., 1*, 1–33 .
16. Cohen, N., Ennaji, H., Hassa, M., & Karib, H., (2006). The bacterial quality of red meat and offal in Casablanca (Morocco). *Mol. Nutr. Food Res., 50*, 557–562. .
17. Dairo, F. A. S., Aina, O. O., & Asafa, A. R., (2005). performance evaluation of growing rabbits fed varying levels of rumen content and blood rumen content mixture. *Niger. J. Anim. Prod., 32*, 67–72.
18. Davidson, A., (2014). *The Oxford Companion to Food* (Vol. 3, pp. 1–29). Oxford University Press: Oxford, UK.
19. Deivasigamani, B., & Alagappan, K. M., (2008). Industrial application of keratinize and soluble proteins from feather keratins. *J. Environ. Biol., 29*, 933–936. .
20. Department of Agriculture Forestry and Fisheries (DAFF), (2011). A Profile of the South African Hides. *Skins and Leather Market Value Chain* (Vol. 1, pp. 1–20). Available online: http://www.daff.gov.za/docs/AMCP/SKINHIDELEAMVCP2011-12.pdf
21. Department of Agriculture, Forestry and Fisheries (DAFF), (2012*). A Profile of the South African Hides, Skins and Leather Market Value Chain* (Vol. 1, pp. 1–22). Available online: http://www.nda.agric.za/docs/amcp/skins2012.pdf (accessed on 30 May, 2020).
22. Deveau, I., Dabbah, R., & Sutton, S., (2004). The USP perspective to minimize the potential risk of TSE infectivity in bovine-derived article used in the manufacture of medical products. *Pharmacop. Forum, 30*, 1911–1921. 88.
23. Elfaki, M. O. A., Abdelatti, K. A., Malik, H. E. E., (2014). Effect of dietary dried rumen content on broiler performance, plasma constituents

and carcass characteristics. *Glob. J. Anim. Sci. Res.*, *3*, 264–270.

24. Enser, M., Hallett, K. G., Hewett, B., Fursey, G. A. J., Wood, J. D., & Harrington, G., (1998). The polyunsaturated fatty acid composition of beef and lamb liver. *Meat Sci.*, *49*, 321–327. .
25. Ercan, P., & El, S. N., (2011). Changes in content of coenzyme Q10 in beef muscle, beef liver and beef heart with cooking and *in vitro* digestion. *J. Food Compos. Anal.*, *24*, 1136–1140. .
26. Falowo, A. B., Muchenje, V., & Hugo, A., (2017). Accepted author version of the manuscript: Effect of sous-vide technique on fatty acid and mineral compositions of beef and liver from bonsmara and non-descript cattle. *Ann. Anim. Sci.*, *17*, 565–580. .
27. Fayemi, P. O., Muchenje, V., Yetim, H., & Abdulatef, A., (2016). Targeting the pains of food insecurity and malnutrition among internally displaced persons with nutrient synergy and analgesics in organ meat. *Food Res. Int., 1, 1–33*. .
28. Florek, M., Litwin´czuk, Z., Skałecki, P., Ke˛dzierska-Matysek, M., & Grodzicki, T., (2012). Chemical composition and inherent properties of offal from calves maintained under two production systems. *Meat Sci.*, *90*, 402–409. .
29. Food and Agriculture Organization of the United Nations (FAO), (2014). *Meat Consumption*. (Vol. 1, pp. 1–20). Available online: http://www.fao.org/ag/againfo/themes/en/meat/background.html (accessed on 30 May, 2020).
30. Gonulalan, Z., Kose, A., & Yetim, H., (2004). Effects of liquid smoke on quality characteristics of Turkish standard smoked beef tongue. *Meat Sci.*, *66*, 165–170. .
31. Gue, D. E., (1998). *Guidelines for Livestock Marketing and Processing Component in Bank Funded Projects* (Vol. 1, pp. 1–33). World Bank: Washington, DC, USA.
32. Hekal, S. A., (2014). Histological study of the skin and leather characteristics in two types of Arabian camels (*Camelus dromedarius*). *J. Am. Sci.*, *10*, 41–48.
33. Hoffman, L. C., Laubscher, L. L., & Leisegang, K., (2013). Nutritional value of cooked offal derived from free-range rams reared in South Africa. *Meat Sci.*, *93*, 696–702. .
34. Hsieh, Y. H. P., & Ofori, J. A., (2011). Blood-derived products for human consumption. *Revel. Sci.*, *1*, 14–21.

35. Hyun, C. K., & Shin, H. K., (2000). Utilization of bovine blood plasma proteins for the production of angiotensin I converting enzyme inhibitory peptides. *Process Biochem., 36*, 65–71. .
36. Irshad, A., & Sharma, B. D., (2015). Abattoir by-product utilization for sustainable meat industry: A review. *J. Anim. Prod. Adv., 6*, 681–696.
37. Jayathilakan, K., Sultana, K., Radhakrishna, K., & Bawa, A. S., (2012). Utilization of by-products and waste materials from meat, poultry and fish processing industries: A review. *J. Food Sci. Technol., 49*, 278–293. .
38. Jones, A. K., Rigby, D., Burton, M., Millman, C., Williams, N. J., Jones, T. R., Wigley, P., et al., (2016). Restaurant cooking trends and increased risk for campylobacter infection. *Emerg. Infect. Dis., 22*, 1208. .
39. Karthikeyan, R., Balaji, S., & Sehgal, P. K., (2007). Industrial applications of keratins: A review. *Am. J. Sci. Ind. Res., 66*, 700–710.
40. Lee, J. W., & Lee, Y. J., (2016). The bacterial quality and prevalence of food borne pathogens of edible offals from slaughterhouses in Gyeongsangbuk-Do. *J. Prev. Vet. Med., 40*, 53–58. .
41. Li, R. R., Yu, Q. L., Han, L., & Cao, H., (2014). Nutritional characteristics and active components in liver from wagyu × qinchuan cattle. *Korean J. Food Sci. Anim. Resour., 32*, 214. .
42. Liu, D. C., (2002). *Better Utilization of* By-Products *from the Meat Industry* (Vol. 1, pp. 1–22). Food and Fertilizer Technology Center: Taipei, Taiwan.
43. Liu, H., Ning, W., Cheng, P., Zhang, J., Wang, Y., & Zhang, C., (2013). Evaluation of animal hairs-based activated carbon for sorption of norfloxacin and acetaminophen by comparing with cattail fiber-based activated carbon. *J. Anal. Appl. Pyrolysis, 101*, 156–165. .
44. Lorenzo, J. M., & Pateiro, M., (2013). Influence of fat content on physico-chemical and oxidative stability of foal liver pâté. *Meat Sci., 95*, 330–335. .
45. Madruga, M. S., Dos, S. M. N., Costa, R. G., De Medeiros, A. N., Do Egypto, R. Q., Schuller, A. R., Albuquerque, C. L. C., et al., (2007). Fat components from precooked "buchada": An edible goat meat by-product. *CYTA J. Food, 5*, 265–270. .
46. Marti, D. L., Johnson, R. J., & Mathews, K. H. Jr., (2012). Where's the (Not) meat? by-products from beef and pork production. *J. Curr. Issues*

*Glob., 4*, 390–397.

47. Masese, L., & Waweru, J., (2017). *Knowledge, Attitudes and Practices Study on Offal Consumption among the Somali Population* (Vol.1, pp. 1–33). Available online: http://www.ennonline.net/fex/41/knowledge (accessed on 30 May, 2020).
48. Meatupdate, (2002). *Storage Life of Meat* (Vol. 1, pp. 1–20). Available online: https://meatupdate.csiro.au/Storage-Life-of-Meat.pdf
49. Miller, J. L., (2013). Iron deficiency anemia: A common and curable disease. *Cold Spring Harb. Perspect. Med., 3*, 011866. .
50. Nakyinsige, K., Man, Y. B. C., & Sazili, A. Q., (2012). Halal authenticity issues in meat and meat products. *Meat Sci., 91*, 207–214. .
51. Nollet, L. M., & Toldrá, F., (2011). *Handbook of Analysis of Edible Animal* By-Products (Vol.1, pp. 1–33). CRC Press: Boca Raton, FL, USA.
52. Ockerman, H. W., & Basu, L., (2004). By-products. In: *Encyclopedia of Meat Sciences* (Vol. 2, pp. 104–112). Elsevier Academic Press: Amsterdam, Netherlands; London, UK.
53. Ofori, J. A., & Hsieh, Y. H. P., (2014). Issues related to the use of blood in food and animal feed. *Crit. Rev. Food Sci. Nutr, 54*, 687–697. .
54. Ørskov, E. R., (2007). Animals in natural interaction with soil, plants, and people in Asia. *Dev. Pract., 17*, 272–278..
55. Osman, A. A., & Elimam, M. E., (2015). Processed animal waste as a feed for Sudanese desert lamb. *Int. J. Adv. Multidiscip. Res., 2*, 12–17.
56. Patrucco, A., Cristofaro, F., Simionati, M., Zoccola, M., Bruni, G., Fassina, L., Visai, L., et al. (2016). Wool fibril sponges with perspective biomedical applications. *Mater. Sci. Eng.61*, 42–50. .
57. Pérez-Alvarez, J. A., Sayas-Barberá, M. E., Sendra, E., & Fernández-López, J., (2011). Color measurements on edible animal by-products and muscle-based foods. In: Nollet, L. M. L., & Toldrá, F., (eds.), *Handbook of Analysis of Edible Animal* By-Products (Vol.1, pp. 87–102). CRC Press: Boca Raton, FL, USA.
58. Phua, S. H., Cullen, N. G., Dodds, K. G., Scobie, D. R., & Bray, A. R., (2016). An ovine quantitative trait locus affecting fiber opacity in wool. *Small Rumin. Res., 130*, 122–126..
59. Pucciarelli, D. L., Friesen, C. A., & Schroeder, A. L., (2012). New culinary uses for pork testicles from immunologically castrated male

pigs. *FASEB J., 26*, 636–646.

60. Purchas, R. W., & Aungsupakorn, R., (1993). Further investigations into the relationship between ultimate pH and tenderness for beef samples from bulls and steers. *Meat Science*, *34*(2), 163–178.
61. Puschner, B., Thurmond, M. C., & Choi, Y. K., (2004). Influence of age and production type on liver copper concentrations in calves. *J. Vet. Diagn. Investig., 16*, 382–387. .
62. Quality Meat Scotland (QMS), (2010). *Recovering Value from the 5th Quarter and Reducing Waste* (Vol.1, pp. 1–19). Available online: www.qmscotland.co.uk/sites/default/files/Added+Value+Topic+3+Edible+Offals.pdf (accessed on 30 May, 2020).
63. Ranken, M. D., (2000). *Handbook of Meat Product Technology* (Vol. 1, pp. 1–30). Blackwell Science: Oxford, UK.
64. Rosa, G. T., Pires, C. C., Da Silva, J. H. S., & Da Motta, O. S., (2002). Proportions and growth coefficients of lamb and lamb carcass non-components in different feeding methods. *Zoo Technical Magazine, 31*, 2290–2298. .
65. Rotenberg, R., (2008). Udders, penises, and testicles. *Ethnology*, *47*, 123–128.
66. Sans, P., & Combris, P., (2015). World meat consumption patterns: An overview of the last fifty years (1961–2011). *Meat Sci.*, *109*, 106–111. .
67. Scobie, D. R., Grosvenor, A. J., Bray, A. R., Tandon, S., Meade, W. J., & Cooper, A. M. B. A., (2015). Review of wool fiber variation across the body of sheep and the effects on wool processing. *Small Rumin. Res., 133*, 43–53. .
68. Seong, P. N., Park, K. M., Cho, S. H., Kang, S. M., Kang, G. H., Park, B. Y., Moon, S. S., & Ba, H. V., (2014). Characterization of edible pork by-products by means of yield and nutritional composition. *Korean J. Food Sci. Anim. Resour., 34*, 290–297. .
69. Sharma, H., Giriprasad, R., & Goswami, M., (2013). Animal fat-processing and its quality control. *J. Food Process. Technol., 4*, 248–252.
70. Sheehy, J., Elmido, A., Centeno, G., & Pablico, P., (2005). Searching for new plants for climate change. *J. Agric. Meteorol., 60*, 463–468. .
71. Steen, L., Glorieux, S., Goemaere, O., Brijs, K., Paelinck, H., Foubert, I., & Fraeye, I., (2016). Functional properties of pork liver protein

fractions. *Food Bioprocess Technol., 9*, 970–980. .

72. Togun, V. A., Farinu, G. O., Ojebiyi, O. O., & Awotunde, A. I., (2010). Effect of replacing maize with a mixture of rumen content and blood meal on the performances of growing rabbits: Initial study with mash feed. *World Rabbit Sci., 17*, 21–26. .
73. Toldrá, F., Aristoy, M. C., Mora, L., & Reig, M., (2012). Innovations in value-addition of edible meat by-products. *Meat Sci., 92*, 290–296. .
74. Tremblay, S., (2011). *The Advantages of Eating Cow Tongue* (Vol. 1, pp.1–20). Available online: https://www.livestrong.com/article/534633-the-advantages-of-eating-cow-tongue/
75. Vanheerden, S. M., & Morey, L., (2014). Nutrient content of South African C2 beef offal. *J. Food Meas. Charact., 8*, 249–258. .
76. Virmond, E., Schacker, R. L., Albrecht, W., Althoff, C. A., De Souza, M., Moreira, R. F., & José, H. J., (2011). Organic solid waste originating from the meat processing industry as an alternative energy source. *Energy, 36*, 3897–3906. .
77. Wismer-Pedersen, J., (1988). Use of hemoglobin in foods: A review. *Meat Sci., 24*, 31–45. .
78. Xiong, G., Han, M., Kang, Z., Zhao, Y., Xu, X., & Zhu, Y., (2016). Evaluation of protein structural changes and water mobility in chicken liver paste batters prepared with plant oil substituting pork back-fat combined with pre-emulsification. *Food Chem., 196*, 388–395. .
79. Yibar, A., Selcuk, O., & Senlik, B., (2015). Major causes of organ/carcass condemnation and financial loss estimation in animals slaughtered at two abattoirs in bursa province, Turkey. *Prev. Vet. Med., 118*, 28–35. .
80. Yitbarek, M. B., Mersso, B. T., & Wosen, A. M., (2016). Effect of dried blood-rumen content mixture (DBRCM) on feed intake, body weight gain, feed conversion ratio and mortality rate of SASSO C44 broiler chicks. *J. Livest. Sci., 7*, 139–149.
81. Zawacka, M., Murawska, D., & Gesek, M., (2016). The effect of age and castration on the growth rate, blood lipid profile, liver histology and feed conversion in green-legged partridge cockerels and capons. *Animal, 11*, 1–10. .

# 4

# *Medical Uses of Animal by-Products*

## CONTENTS

## 4.1. INTRODUCTION

It is known to all that in animal-based medicinal products, the annual global trades contribute to billions of dollars per annum. In the development of bioprospecting, the examination of conventional medicines has demonstrated value for the pharmaceutical compounds. Around 11.1% of the 252 critical chemicals come from plants, and 8.7% from animals that have been chosen by the World Health Organization (WHO) (Marques, 1997). Around 27 prescription drugs have animal origin out of 150 which are presently in use in the United States of America (World Resources Institute, 2000) (Figure 4.1).

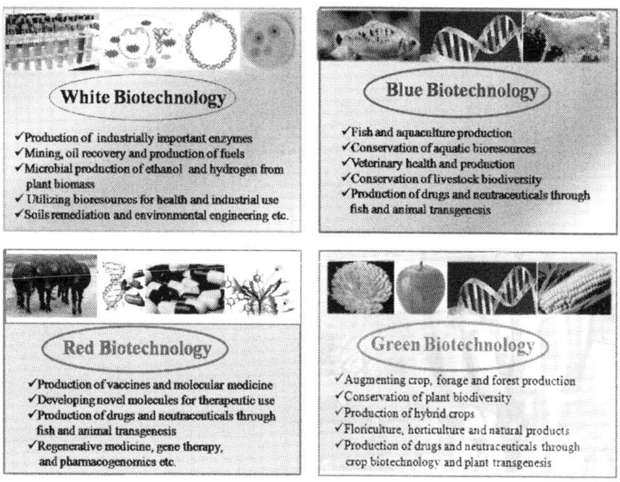


**Figure 4.1:** Miscellaneous uses of animal and plant-based By-Products for production of medicines.

*Source: https://link.springer.com/chapter/10.1007/978-3-030-21309-1_48.*

In this chapter, the phenomenon of zoo therapy will be discussed briefly, to stress how significant animals were, are, and will be as foundations of pharmacological substances. Knowledge of clinical and traditional use of medicinal animals can be enhanced by filling in blanks about these topics in this chapter. Moreover, the significance of animals as the origin of the potential drugs has been presented in this chapter as well as a critical and wider discussion on the sustainable use of species and conservation biology has also been presented.

## 4.2. THE USE OF ANIMALS IN TRADITIONAL MEDICAL SYSTEMS

Throughout the world, the conventional medical knowledge of native peoples has played a significant role in recognizing biological assets worthy of commercial utilization. Indeed, ethno biological data has guided the search for novel pharmaceuticals from naturally occurring organic material (Blakeney, 1999). For instance, the medicinal use of 50 animals has been recorded by the Alexiades (unpublished data) by the Ese Eja people from Peru. To treat the stroke and epilepsy the blood of the black caiman (*Melanosuchus niger* [Spix1825]) was used and for toothache ants of the genus, *Pseudomyrmex* is smashed or are used to bite painful joints. There is a famous quote in which people of the area of Sierra Madre used to say that is "The more poisonous the animal, the more potent its anti-poison" (Werner, 1970). For infirmities ranging from bronchitis to boils, numerous anatomical parts of the rattlesnake (*Crotalus* sp.) have been used as pointed out by this author.

In the Sudanese conventional medicine, 23 animals have been recorded by El-Kamali (2000) that are used as sources of remedies. For instance, on the affected parts to alleviate arthritis, the fresh manure of a dromedary (*Camelus dromedaries* [Linnaeus 1758]) is applied externally; gastrointestinal and hepatic disorders, gastric ulcers and also to heal wounds the honey has been used; To alleviate abdominal pains, the fats of the hyena (*Crocuta crocuta* [Erxleben 1777]) and lion (*Panthera leo* [Linnaeus 1758]) are used. 23 species have also been recorded by Adeola (1992) in three ecological zones from Nigeria. He signifies for the preparation of traditional medicine and daily animal protein supply that most of the farmers in the rural areas of Nigeria rely only on wild animals and their by-products (bones, hooves, skins, tusks, feathers). For instance, for ornamentals and aphrodisiacs, the tusks of hippo (*Hippopotamus amphibious* [ Linnaeus 1758]) are used. To cure rheumatism, backache, and boils, fat extracted from a manatee (*Trichechus senegalensis* Link, 1795) is used. To appease or invoke witches and traditional gods, special vessels for concoctions with herbs are mainly hollow parts of the hooves of antelopes (*Hippotragus equines* [Desmarest 1804]) and duikers (*Sylvicapra grimmie* [Linnaeus 1758]). According to most of the Africans belief, special healing acts are attached to some magical powers when they use by-products of the wild animals as prescribed by the conventional healer. Hence, in the preparation of drugs, conventional medicine man employs various ways which include mammals, herbs, birds, bark, leaves, and roots. Conventional medicine remains the major or only source of health care for the majority of people in both Nigeria and Sudan. Though, wild animals hunting for medicinal and meat purposes is a vital issue in all national parks and game reserves from Africa.

Medicinal and research uses of earthworms have an origin of nearly 4,000 years in China (Zhang et al., 1992). A comprehensive summary of pharmacological knowledge has been given in *Compendium of Materia Medica* written by Li Schizhen in 1578 AD accumulated in China up to his time. Earthworms have diuretic, detoxic, antiallergic, antispasmodic, spermicidal, antipyretic, antihypertensive, and antiasthmatic effects according to conventional Chinese medicine. Over 80 diseases can be treated by the prescription of earthworm medicines (e.g., mumps, cancer, asthma, epilepsy, hypertension, ulcer, etc.). Further study is required on earthworm extracts especially as a new spermatocide (Zhang et al., 1992).

To cure both complicated and common ailments, over 500 species of mites, insects, and spiders are used as medicine in Chhattisgarh, India (Oudhia, 1995). For instance, for paralysis the red velvet mite (*Trombidium grandissimum* Koch, 1867) oil is useful. These mites are called Indian Viagra, because of its ability to boost sexual desires. More than 50 common diseases can be cured by the pod borer *Helicoverpa armigera* (Hübner, 1805) alone or in mixture with herbal drugs. Through feeding habits and particular behavior, the identification of high medicinal value insects is easy as described by the folk doctors from Chhattisgarh. Indeed, the discovery of useful compounds can be helpful by the behavior of insects by helping the observer to an unusual chemical (Joyce, 1991). Chemistry is at work if predators avoid an insect's egg when it is covered with its mother's saliva, or ants turn up their noses at a fallen leaf. Ecologists can spot interesting compounds, by keeping a lookout for this kind of telltale behavior.

Various marine/estuarine animal resources have been in use by Artisanal fishermen from Siribinha Beach in the State of Bahia, Northeastern Brazil, as folk medicines (Costa-Neto and Marques, 2000). When questioned about their folk medicine, 24 fish species were recorded as having some therapeutic use. 15 other animals having medicinal features were also cited, although interviews focused on fish-based remedies. So, there are a total of 39 resources, which are divided into six taxonomic categories, i.e., mammals (2%), mollusks (5%), echinoderms (8%), reptiles (10%), crustaceans (13%), and fish (62%). To treat locally diagnosed ailments in the elaboration of remedies, a total of 66 raw materials which include spur, fat, scales, skin,

besides hunting to be sold, after they are grown, because of their bladder, skin, and meat (Balzar, 1992).

Although in Brazil, the slaughtering, hunting, and trading of wild animals have been restricted by the federal law since 1967, they still use both medicinally and nutritionally in a clandestine way. Around 34 animals have been investigated by Oliveira and Costa-Neto (2000) which are thought to be cosmetics, folk medicines, and charms in the county of Tanquinho, northeastern of the State of Bahia. Around 24 of the total species recorded are not under the risk of extinction. Moreover, *Dusicyon* sp., *prehensile*, *Coendou* cf., *Crypturellus noctivagus Zabeel*, *Americana*, *tridactyla*, *Rhea americana*, Myrmecophaga, and *Mazama* cf. which have been declared officially threatened species by the Brazilian Institute of Environment and Renewable Natural Resources, were found among the set of faunistic assets recommended as medicines at the time of this research (Rocha, 1995; Da Silva et al., 2008). At least three species are not sufficiently known and hence they are considered to be threatened. These include tortoise, *T. pecari* and *Tayassu tajacu.* Lastly, *Phrynops* is a less known genus that is supposed to include threatened species (Grelle et al., 2005; Mascarenhas et al., 2009)

Worldwide, quickly expanding trade for conventional medicines, aquarium fishes, aphrodisiacs, foods, and curious directly threaten the seahorses and their pipefish relatives and the destruction of their seagrass, coral reef habitats, and mangrove indirectly threatens these species (Vincent, 1999). Since 1990 in Indonesia the population of some of the species has decreased to half, with pregnant males being the most normal prey (Poléjaeff, 1884; Faulkner, 1992; Carté, 1996; Fusetani, 1996).

## 4.4. ZOOTHERAPY AND ITS SUSTAINABILITY

Only in the State of Bahia, the record of over 180 medicinal animals as well as the research conducted in foreign countries and other parts of Brazil, show a powerful indication of the medicinal use of wildlife resources (Costa-Neto 2004). Oldfield (1989) discusses that various animal species have been overexploited as a cause of medicines for the folk medicine trade. Moreover, she also claims that the population of animals has become endangered or depleted as a consequence of their use as animal models or experimental subjects. Sustainability is now essential for this reason as the administrative principle for biological preservation. The exploitation of a given species is likely to be sustainable According to the IUCN draft *Guidelines* (Glowka et al., 1994), if (Vilar et al., 2011; Agudo-Padrón, 2015):

1. It does not decrease the future use of the potential of the objective population or damage its long-term feasibility;
2. It is well-suited with the conservation of the long-term feasibility of supportive and dependent ecosystems; and
3. It does not impair the long-term viability of other species or reduce the future use potential.

But weather properly managed, the zootherapeutic activity can be well-matched with an environmental preservation program in which the use of natural assets can and must happen in such a way that protection of biodiversity and human needs be guaranteed (Andriguetto-Filho et al., 1998). Zoo therapy should be observed within its cultural dimension, for this reason (Costa-Neto 1999a).

How people perceive, allocate, transfer, use, and manage their natural assets is included in this cultural perspective (Johannes, 1993). Suppression of the animal use will not save them from extinction since people have been using them for a long time.

Hence, those species which have been directly involved in the conventional medicines should be in the list of the highest priorities for the conservation, under Kunin et al. (1996).

Some of these species are endangered precisely as attested by these authors since they are valuable to us.

The extraction rate of a renewable resource should not increase above the renovation rate of the same resource since it is the basic principle governing the use natural resources, the localization of natural compounds would perhaps be an appropriate substitute for the impoverishment of wild resources overexploitation that has been positively verified for pharmacological action. Hence, human dependency on animal medicines would be displaced by the production of artificial substitutes in the laboratories (Oldfield, 1989; Jonathan, 2012).

Turning the endangered species into manageable resources in the means of conventional farming systems is another alternative for the recovery of endangered species, where they would be raised using both scientific and folk methods (Carte, 1996; Costa-Neto 1999b).

## 4.5. ANIMALS PHARMACOLOGICAL IMPORTANCE

Pharmaceutical companies have tested animals methodically as sources of drugs to modern medical science (Gurib-Fakim, 2006). The inhibitors of angiotensin in converting enzyme (ACE) is the best-known example of effective drug development from an element of snake venom (*Bothrops jararaca*). Conversion of the inactive precursor into the locally active hormone angiotensin is the responsibility of the enzyme, which results in blood vessels to narrow and hence raises blood pressure (Bisset, 1991). Blood pressure can be reduced by pharmacologically blocking the enzyme's activity and commonly agents having this activity are used to treat people with high blood pressure (Harvey, 1995; Francis, 1996). Presently, ACE inhibitors such as lisinopril, enalapril, and captopril are in the top twenty bestselling medicines in the world (Pallas, 1771; Phillipson, 1989; Jones, 1998; Konno et al., 1998) (Figure 4.3).

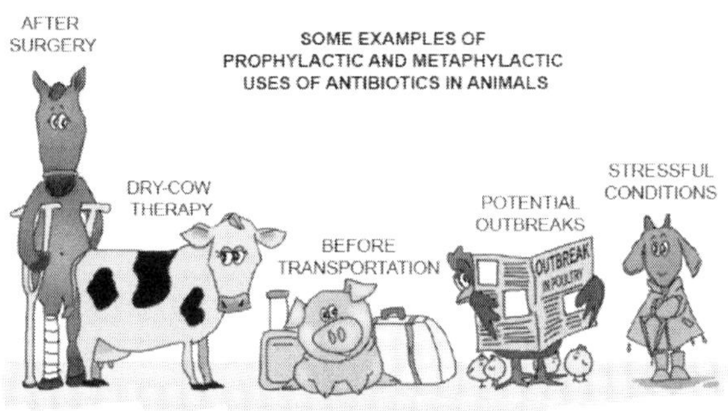


**Figure 4.3:** Pharmacological importance of animals.

*Source: http://amrls.cvm.msu.edu/pharmacology/antimicrobial-usage-in-animals/non-therapuetic-use-of-antimicrobials-in-animals/use-of-antibiotics-in-animals-for-prophylactic-or-metaphylactic-purposes.*

Numerous compounds have been extracted regarding fish and are used as remedies in official medicine (Hamada and Nagai, 1995). For biochemical research today, some of these compounds act as an important tool or as new areas for the development of antiviral and anticancer drugs. A new chemical compound derived from dogfish sharks (*Squalus acanthias* Linnaeus, 1758) is one of the most promising discoveries of the 1990s, which kills fungi, parasites, and bacteria as stated by Agosta (1996). In wounded sharks, this chemical aid prevent infection and someday it is expected to do the same for humans. Finkl (1984), for instance, refers to *Dasyatis sabina* (Lesueur, 1824), *Eptatretus stout* (Lockington, 1878), and *Taricha* sp. as a source of antitumors, stimulants, and analgesic, respectively. Oily fish, like herring, turbot, cod, and salmon, have great potential; as medicinal value to human beings because the polyunsaturated compound is known as OMEGA-3 (Beattie1992; Young, 1999). The prevention of arthritis is helped by this substance (Adeodato 1997). The existence of an anticoagulant system in the plasma of rainbow trout (*Oncorhynchus mykiss* [Walbaun 1792]) and Atlantic salmon (*Salmo salar* Linnaeus, 1758) has been verified, what helps similarities with the protein C anticoagulant system in mammals (Salte et al., 1996). A bioactive compound tetrodotoxin (TTX) which is soluble in water guanidinium derivative is produced by marine organisms such as pufferfish "it is similar to procaine in its ability to inhibit transmission of nerve cells" (Colwell, 1997). It acts as an extraordinary narcotic and analgesic when diluted (Bisset, 1991).A type of frog is being studied by the Brazilian scientists which are used to intestinal illnesses by associates of the Yawanawa Indian communities on the banks of the Rio Grande, according to McGirk (1998). Indeed, compounds having the ability to be turned to therapeutic advantage have been provided by the amphibians. For instance, in the treatment of depression, seizures, stroke cognitive loss in ailments such as Alzheimer's disease, peptides extracted from the scraped secretions of *Phyllomedusa bicolor* (Boddaert, 1772) are used (Amato 1992).

The diversity and number of compounds produced by amphibians in their granular glands are astonishingly high, even within single species. The following are the main categories of the secretion including bufotoxins and bufotenine (steroids), biogenic amines alkaloids and peptides (including polypeptides, smaller oligopeptides, and proteins) and the as yet uncharacterized zetekitoxins found in the South and Middle American harlequin frogs, genus *Atelopus*. From dietary arthropods, most alkaloids of amphibian skin seem to be sequestered (Daly, 1998). The defensive role is played by many of the compounds in the granular gland secretions. Neurotoxic, myotoxic, and cardiotoxic activities are included in their pharmacological effects, some are hypotensive and vasoconstrictive agents while others have hallucinating effects-all the features which would affect the potential hunter. A novel opioid heptapeptide known as dermmophin produced in the skin of the South American leaf frogs, genus *Phyllomedusa*, has a 1000 times more effective than morphine at the same dosage level (DeFoliart, 1992; Clarke, 1997).It is hence possible that molecules derivatives of the amphibian might deliver useful substitute or supplementary treatments chiefly as antimicrobial (including antifungal) agents. The existence of a wide range of peptides with unknown features increases the projections of using peptide vaccines for viral diseases (Clarke, 1997; Omotoso, 2006).

## REFERENCES


1. Adeodato, S. E. R. G. I. O., (1997). Os santos remédios do mar. *Globo Ciência.*, *4*, 20–25.
2. Adeola, M. O., (1992). Importance of wild animals and their parts in the culture, religious festivals, and traditional medicine, of Nigeria. *Environmental Conservation*, *19*(2), 125–134.
3. Agosta, W. C., (1996). *Bombardier Beetles and Fever Trees: A Close-Up Look at Chemical Warfare and Signals in Animals and Plants,* (Vol. 1, pp.1–33).
4. Agudo-Padrón, A. I., (2015). Balance of the Brazilian *Molluscs* officially recognized as threatened of extinction, with special emphasis in species occurring in the Southern Region. *Brazilian Journal of Biological Sciences*, *2*(3), 173–175.
5. Amato, I., (1992). From 'hunter magic,'a pharmacopeia? *Science*, *258*(5086), 1306–1307.
6. Andary, C., Motte, E., Elorduy, J., Privat, A., & Perez-Ramos, J., (1996). *Chemical Screening: Updated Methodology Applied to Medicinal Insects* (Vol. 1, pp. 1–44.
7. Andriguetto-Filho, J. M., Krüger, A. C., & Lange, M. B. R., (1998). Hunting, Biodiversity and environmental management in the Environmental Protection Area of Guaraqueçaba, Paraná, Brazil (Vol. 1, pp. 1–30). Biotemas.
8. Balzar, J., (1992). Uma cultura antiecológica. *Jornal. Do Brasil*, *22, 1–30.*
9. Beattie, A. J., (1992). Discovering new biological resources: Chance or reason? *BioScience*, *42*(4), 290–292.
10. Bisset, N. G., (1991). One man's poison, another man's medicine? *Journal of Ethnopharmacology*, *32*(1–3), 71–81.
11. Blakeney, M., (1999). What is traditional knowledge? Why should it be protected? Who should protect it? For whom?: Understanding the value chain. World Intellectual Property Organization (WIPO) Roundtable on intellectual Property and Traditional Knowledge, WIPOIIPTKIRTI9913, 1–40.
12. Blouin, M., (1991). Rhino horns and humans. *Zoo Goer.*, *1*, 1–20.
13. Carte, B. K., (1996). Biomedical potential of marine natural products. *Bioscience*, *46*(4), 271–286.

14. Clarke, B. T., (1997). The natural history of amphibian skin secretions, their normal functioning and potential medical applications. *Biological Reviews*, *72*(3), 365–379.
15. Colwell, R. R., (1997). Microbial biodiversity and biotechnology. In: Reaka-Kudla, M. L., Wilson, D. E., & Wilson, E., (eds.), *Biodiversity II: Understanding and Protecting our Biological Resources* (Vol. 1, pp.1–30). Washinton, DC.
16. Costa-Neto, E. M., & Oliveira, M. V. M., (2000). Cockroach is good for asthma: Zoo therapeutic practices in Northeastern Brazil. *Human Ecology Review, 1*, 41–51.
17. Costa-Neto, E. M., (1999a). Healing with animals in Feira de Santana city, Bahia, Brazil. *Journal of Ethnopharmacology*, *65*(3), 225–230.
18. Costa-Neto, E. M., (1999b). Animal resources used in the traditional medicine of the Pankararé Indians who live in the northeast of the state of Bahia, Brazil. *Biological News, 21(*70), 69–79.
19. Costa-Neto, E. M., (2004). Implications and applications of folk zoo therapy in the state of Bahia, Northeastern Brazil. *Sustainable Development*, *12*(3), 161–174.
20. Da Silva, A. P., & Pontes, A. R. M., (2008). The effect of a mega-fragmentation process on large mammal assemblages in the highly-threatened pernambuco endemism centre, north-eastern Brazil. *Biodiversity and Conservation*, *17*(6), 1455–1464.
21. Daly, J. W., (1998). Thirty years of discovering arthropod alkaloids in amphibian skin. *Journal of Natural Products*, *61*(1), 162–172.
22. De Salltalla, F., (2000). Faunistic resources used as medicines by artisanal fishermen from Siribinha beach, state of Bahia, Brazil.' *Journal of Ethnobiology*, *20*(1), 93–109.
23. DeFoliart, G. R., (1992). Insects as human food: Gene De Foliart discusses some nutritional and economic aspects. *Crop Protection*, *11*(5), 395–399.
24. DEPREkEL, M., (2002). *Applications of Animal Assisted Therapy, 1*, 1–30.
25. Diamond, G., (2001). the potential of antimicrobial peptides as new drugs. *Biologist*, *48*(5), 1–40.
26. Ekop, E. A., Udoh, A. I., & Akpan, P. E., (2010). Proximate and anti-nutrient composition of four edible insects in Akwa Ibom State, Nigeria. *World J. Appl. Sci. Technol*, *2*(2), 224–231.

27. El-Kamali, H. H., (2000). Folk medicinal use of some animal products in Central Sudan. *Journal of Ethnopharmacology*, *72*(1/2), 279–282.
28. Francis, D. G., (1996). Equoterapia: Recurso inovador para reabilitação física e mental. *Anais De Etologia*, *14*, 59–63.
29. Fusetani, N., (1996). Bioactive substances from marine sponges. *Journal of Toxicology: Toxin Reviews*, *15*(2), 157–170.
30. Glowka, L., Burhenne-Guilmin, F., & Synge, H., (1994). *Guide to the Convention on Biological Diversity* (Vol. 1, pp.1–22). IUCN--the World Conservation Union.
31. Grelle, C. E. V., Alves, M. A. S., Bergallo, H. G., Geise, L., Rocha, C. F. D., Van, S. M., & Caramaschi, U., (2005). Prediction of threatened tetra pods based on the species-area relationship in Atlantic Forest, Brazil. *Journal of Zoology*, *265*(4), 359–364.
32. Gurib-Fakim, A., (2006). Medicinal plants: Traditions of yesterday and drugs of tomorrow. *Molecular aspects of Medicine*, *27*(1), 1–93.
33. Hamada, M., & Nagai, T., (1995). Inorganic components of bones of fish and their advanced utilization. *Journal of Shimonoseki University of Fisheries*, *43*(4), 185–194.
34. Harvey, A. L., (1995). From venoms to toxins to drugs. *Chemistry and Industry*, (22), 914–916.
35. Higa, T., & Kuniyoshi, M., (2000). Toxins associated with medicinal and edible seaweeds. *Journal of Toxicology: Toxin Reviews*, *19*(2), 119–137.
36. Huxtable, R. J., (1992). The pharmacology of extinction. *Journal of Ethnopharmacology*, *37*(1), 1–11.
37. Johannes, R. E., (1993). Integrating traditional ecological knowledge and management with environmental impact assessment. *Traditional Ecological Knowledge: Concepts and Cases*, *1*, 33–39.
38. Jonathan, A. A., (2012). Proximate and anti-nutritional composition of two common edible insects: Yam beetle (*Heteroligus meles*) and palm weevil (*Rhynchophorus phoenicis*). *Food Sci.*, *49*, 9782–9786.
39. Jones, D., (1998). The neglected saliva: Medically important toxins in the saliva of human lice. *Parasitology*, *116*(S1), S73–S81.
40. Joyce, C., (1991). Prospectors for tropical medicines. Descubridores de medicinas tropicales. *New Scientist.*, *132*(1791), 36–40.
41. Konno, K., Hisada, M., Itagaki, Y., Naoki, H., Kawai, N., Miwa, A.,

& Takayama, H., (1998). Isolation and structure of pompilidotoxins, novel peptide neurotoxins in solitary wasp venoms. *Biochemical and Biophysical Research Communications, 250*(3), 612–616.

42. Kunin, W., Lawton, J. H., Kunin, A. A., & Lawton, H., (1996). Does biodiversity matter? *Evaluating the Case for Conserving Species, 1*, 1–33.
43. Launet, E., (1993). In the woods, looking for the medicines of tomorrow. *Science and Life, 904* , 86-91.
44. Lazarus, L. H., & Attila, M., (1993). The toad, ugly and venomous, wears yet a precious jewel in his skin. *Progress in Neurobiology, 41*(4), 473–507.
45. Marks, J. G. W., (1994). The medicinal fauna of the Kuna Indians of San Blas (Panama) and the hypothesis of zootherapeutic universality. *Annals of the 46th Annual Meeting of the SBPC* (Vol. 46, pp. 300–324).
46. Marques, J. G., (1997). Fauna medicinal: Recurso do ambiente ou ameaça à biodiversidade. *Mutum, 1*(1), 4.
47. Martin, E. B., (1992). Observations on wildlife trade in Vietnam. *TRAFFIC Bulletin, 13*(2), 61–67.
48. Mascarenhas, C. S., Krüger, C., & Müller, G., (2009). The helminth fauna of the red-crested cardinal (Paroaria coronata) Passeriformes: Emberizidae in Brazil. *Parasitology Research, 105*(5), 1359–1363.
49. McGirk, T., (1998). Gene piracy. *Time, 30*, 20–25.
50. Midgley, M., (1993). "The four-leggeds, the two-leggeds, and the wingeds": An overview of society and animals. *Society and Animals, 1*(1), 9–15.
51. Oldfield, M. L., (1984). *The Value of Conserving Genetic Resources* (Vol. 1, pp. 1–20) US Department of the Interior, National Park Service.
52. Omotoso, O. T., (2006). Nutritional quality, functional properties and anti-nutrient compositions of the larva of Cirina forda (Westwood) (Lepidoptera: Saturniidae). *Journal of Zhejiang University Science, B., 7*(1), 51–55.
53. Oudhia, P., (1995). Traditional knowledge about medicinal insects, mites and spiders in Chhattisgarh, India. *Insect Environment, 4*, 57–58.
54. Pentenero, L., (2001). *Los Animales Ayudan a Afrontar Problemas* (Vol. 24, pp. 1–30). La Nación. Bogotá.
55. Phillipson, J. D., (1989). Natural products as pharmacological probes

and in new pharmaceuticals. *The Linnean*, *5*, 29–33.

56. Rocha, C. F. D., (1995). Nematode parasites of the Brazilian sand lizard, *Liolaemus lutzae*. *Amphibia-Reptilia*, *16*(4), 412–415.
57. Salte, R., Norberg, K., & Ødegaard, O. R., (1996). Evidence of a protein C-like anticoagulant system in bony fish. *Thrombosis Research*, *83*(5), 389–397.
58. Shelomi, M., (2015). Why we still don't eat insects: Assessing entomophagy promotion through a diffusion of innovations framework. *Trends in Food Science and Technology*, *45*(2), 311–318.
59. Unnikrishnan, P. M., (1998). Animals in ayurveda. *Amruth*, *1*(3), 1–23.
60. Vilar, C. C., Spach, H. L., & Santos, L. O., (2011). Fish fauna of Baía da Babitonga (southern Brazil), with remarks on species abundance, ontogenic stage and conservation status. *Zootaxa 2734*(1), 40–52.
61. Vincent, A., (1999). *Australia and the International Trade in Seahorses* (Vol. 1, pp. 1–22). Available at: http://www.amcs.org.au/ periodic/bulletin/bull203/bull203.htm (accessed on 30 May, 2020).
62. Von, M. K. F. P. V., (1939). *Nature, Diseases, Medicine and Remedies of Brazilian Indians* (Vol. 1, pp. 1–30). *Brasiliana*.
63. Werner, D., (1970). Healing in the sierra Madre. *Natural History*, *79*(9), 61–66.
64. World Resources Institute, (2000). *World Resources: People and Ecosystems: The fraying Web of Life* (Vol. 2000, pp. 1–50). World Resources Institute.
65. Yamakawa, M., (1998). Insect antibacterial proteins. *The Journal of Seri Cultural Science of JAPAN*, *67*(3), 163–182.
66. Young, R. N., (1999). Importance of biodiversity to the modern pharmaceutical industry. *Pure and Applied Chemistry*, *71*(9), 1655–1661.

# 5

# *Application of Plant by-Products in Cosmetics*

## CONTENTS

## 5.1. INTRODUCTION

An essential part of our diet is represented by vegetables and fruits and for decades their undeniable worth has been extensively studied. Other than primary metabolites (e.g., lipids, saccharides, and amino acids), the plants manufacture a majority range of "secondary" metabolites, which have a vital role in the protection of plants against both abiotic and biotic stresses (Barbulova et al., 2014). Furthermore, various types of interactions with the environment and with other organisms that lodge the same environmental habitat have been medicated by the secondary metabolites (Oksman-Caldentey and Inze, 2004). As fragrances, preservatives, dyes, pharmaceuticals, and flavors, they have been of great interest for centuries: consumption of vegetables and fruits has been indicated by the epidemiological researches which shows the great benefits of health, such as certain types of cancer as well as reduced risk for stroke and coronary heart disease. The effects of secondary metabolites and dietary fibers are the main protective action, mainly flavonoids, and carotenoids with powerful antioxidant activity. The most copious carotenoid, beta-carotene, was proficient in neutralizing the free radicals contained in red and orange fruits such as tomatoes and growing the cellular antioxidant defenses. The most vital examples of mixtures with therapeutic activity among the flavonoids are epicatechins, catechins, procyanidins, and flavanols, present in grapes and apples, that subsidize to preserve the heart health; the flavanones, the foremost flavonoid type in citrus fruits; the anthocyanidins, mostly present in red grapes, berries, cherries, that enhance the brain functions and rise the antioxidant defenses (Kris-Etherton et al., 2004; Djilas et al., 2009) (Figure 5.1).

**Figure 5.1:** Image showing different plant-basedcosmetics.

*Source: https://seybeautenaturelle.com/.*

Thanks to all these features and to the wide range campaign to encourage the benefits of fresh green products, in the last decades, the consumption of vegetables and fruit have enhanced substantially and the residues and wastes, are now estimated to be around 800,000 tons per year of fresh matter globally, produced by agro-food industries, without taking into account the processing of food (Ayala-Zavala et al., 2004).

Today, valorization strategies and basic waste management techniques are used for the processing of the agricultural waste as the production of fertilizers and composts, animal feed as well as anaerobically digestion to biogas.

Environment-friendly and efficient recycling and utilization of this waste are becoming more and more critical since this kind of processing is linked with the high costs of transport, storage, and drying (Tuck et al., 2012).

10~60% of the raw materials are generated by the agricultural processing industries as solid waste and it was it has been proved that in some scenarios the waste by-product was even more useful than the main products (Goñi et al., 2011).

Vegetable and fruit processing industries produce waste materials which have been illustrated in Table 6.1 expressed as percentages, which we wish to call "by-products" to emphasize their large potentialities to be recycled as appreciated products for industry. Seeds, leaves, skins, stems, unstable pulp and wastewaters are the by-products which in some scenarios represent more than 40% of total food (such as the cases of asparagus, papaya, artichoke, pineapple, citrus fruits, mango) and are normally discarded.

It is a known fact that by-products of food are very rich in minerals, dietary fiber sugars, organic acids, bioactive compounds, as carotenoids and polyphenols, analogously to their edible counterparts.

Their attracting composition along with the increasing interest in searching natural constituents as a substitute to synthetic substances has found that the by-products of food are the economically beneficial source for the synthesis of high-value compounds in various fields of industry which include

Pharmaceutics, Food, Cosmetics, and Nutraceutics (Schieber et al., 2001). This review covers some examples among the most significant application of agricultural by-products of food in Cosmetics, and argue their performance as safety and efficacy (Table 5.1).

**Table 5.1**: By-Product Quantities Generated from Some Fruit and Vegetable Processing Industries

| Plant | By-Products | Edible Part |
|---|---|---|
| Agave | 40% (rind and pith) | 60% |
| Citrus fruits | 66% (peel) | 44% |
| Pineapple | 52% (core, peels, top, pulp) | 48% |
| Apple | 11% (pulp and seed core) | 89% |
| Artichoke | 60% (outer bracts, receptacles, and stems) | 40% |
| Cactus cladodes | 20% (spines, glochids, and peel) | 80% |
| Carrot | 30%–40% (pomace) | 60%–70% |
| Banana | Up to 30% (peel) | 70% |
| Asparagus | Up to 40%–50% (spear) | 50%–60% |
| Mango | 42% (seeds, peels, unusable pulp) | 58% |
| Tomato | 3%–7% (peel and seeds) | 93%–97% |
| Papaya | 47% (seeds, peels, unusable pulp) | 53% |
| Passion fruit | 75% (rind and seeds) | 25% |
| Potato | 15%–40% (peel) | 60%–85% |
| Mandarin | 16% (peels) | 84% |

## 5.2. CITRUS FRUIT PROCESSING BY-PRODUCTS

One of the world's main fruit crops are citrusplants (genus *Citrus* in family *Rutaceae*), also recognized as *agrumes*, with global popularity and availability contributing to the human diet. One of the sources of dietary fibers are fresh citrus fruits connected with lowered circulating cholesterol and gastrointestinal disease prevention; they are rich in vitamin B, C vitamins (pantothenic acid, niacin, folate, pyridoxine, riboflavin, and thiamin), and phytochemicals, i.e., limonoids, flavonoids, and carotenoids (Liu et al., 2012; Kim et al., 2013). In human health development, the biological ingredients are of vital importance because of their ability to be changed to vitamin A (for instance the β-cryptoxanthin), antioxidant properties, and to shelter from numerous chronic diseases. About one-third of the citrus fruits are utilized after processing worldwide although many citrus fruits, such as grapefruits, tangerines, and oranges can be eaten fresh, and lemon/orange juice production is considered for nearly 85% of total processed intake (Figure 5.2).

**Figure 5.2:** Cosmetic products produced from citrus fruits.

*Source: https://manofmany. com/fashion/mens-fragrances/15-best-fresh-citrus-fruity-colognes.*

High quality and high value of agricultural waste are generated by citrus processing, and most of the waste is in the shape of press cakes, that due to costs for transport and handling are pricy to dispose of. A study was carried out into the anti-melanogenic impacts of ethanol extracts of the press cakes of *Citrus unshiu*, to give value to this waste, the fruit peel was widely used to treat atopic and severe dermatitis or in the conventional medicine as digestive, which were mainly grown in Jeju Island in Korea (Kim et al., 2008).

From natural sources, the development of novel whitening phytochemicals has become a very common trend. The color of human eyes, hair, and skin are determined by the end product of melanogenesis known as Melanin and although it has some limitations such as accumulation and overproduction in the skin may cause aesthetic problems and skin disorders and problems like dark spots and hyper-pigmentation due to its photoprotective role in absorbing free radicals in the cells and shielding from UV light (Balkrishnan et al., 2006; Costin and Hearing, 2007).

The transcription factor MITF (Microphthalmia-associated transcription factor and the activity of the main enzyme, tyrosinase determine the production of Melanin; hence, reduced melanin content in the skin would be caused by the expression or a down-regulation tyrosinase activity. Moreover, for melanin synthesis, tyrosinase-related protein TRP-2 and TRP-1 are also vital and are regulated by MITF (Yen et al., 2012).

Using ethanol as solvent, citrus press cakes composed of *C. unshiu* fruit peels were extracted its capacity to inhibit protein expression profile of TRP-

2 and TRP-1 and MITF, tyrosinase activity and melanin content in murine B16F10 melanocytes was tested. It was determined that the cellular melanin content was significantly reduced by the treatment with citrus press cake extracts, through the transcription factors TRP-2 and TRP-1 and inhibition of the tyrosinase activity in a dose-dependent manner. Based on the Western blot examination (analytical method, used to notice precise proteins in a sample), in a dose-dependent manner, the upstream transcription factor MITF was also down-regulated.

These results recommended that this kind of by-products of food was an auspicious applicant for treating skin pigmentation disorders, which should be complemented with further *in vivo* valuation of the biological activity of citrus press cake extracts (Sabio et al., 200; Papaioannou and Karabelas, 2012).

Citrus by-products, besides the ethanol extracts, are the significant basis of essential oils (Pavithra et al., 2009). Because of their strong antioxidant, antimicrobial, and anti-inflammatory features, essential oils acquired from plants are well known and have a large number of suitable applications which include use as preservatives against spoilage, food additives, cosmeceuticals, and pharmaceuticals (Baik et al., 2008). Hence, the study was done into the potential utilization of an essential oil using hydro-distillation obtained from *C. unshiu* peel wastes, as an anti-microbial and anti-inflammatory agent in Cosmetics (Yang et al., 2009). The existence of six main compounds was revealed in the chemical analysis of this oil (containing up to 94.5% of the total oil composition): and α-terpinolene (0.56%), α-pinene (1.02%), β-myrcene (1.59%), cymene (4.02%), γ-terpinene (6.80%), limonene (80.5%). The oil due to its monoterpene composition displayed a strong antibacterial activity, which inhibits the respiration process damaged the cellular integrity of the microbial cells. It was displayed that nitric oxide (NO) synthesis can be down-regulated by the oil. NO is synthesized in many animal tissues and cells and is an endogenous free radical species produced by nitric oxide synthase (NOS), and also the inflammatory process is activated by the oils in the tissues. NO synthesis can be induced in numerous cells, after exposure to stress inducers, i.e., bacterial lipopolysaccharide (LPS), triggering a transduction cascade causing inflammation and tissue damage (Muracami, 2009). *C. until* the oil, produced a strong inhibition of NO production extracted from the press-cakes, when tested on LPS-treated RAW264.7 cell line (macrophages of the immune system), recommending that it could be beneficial anti-inflammatory component to be used in cosmetic formulas (Zuorro et al., 2014; Yilmaz et al., 2016).

## 5.3. TOMATO AND OLIVE PROCESSING BY-PRODUCTS

The olive oil production farms and the tomato processing industries are the two most significant processes which produce valuable products exported worldwide, these are mostly located in the Mediterranean areas.

Approximately 160 million tons of the fresh tomatoes are produced annually on a global scale and about 40 million tons of tomatoes are processed as tomato products: chopped or whole, unpeeled or unpeeled tomatoes, tomato paste (Strati and Oreopoulou, 2014; Barbulova et al., 2015). A large number of fascinating by-products are produced by the tomato processing, seeds, and pulps, also known as "tomato pomace," peels, which are significantly rich in the antioxidant compounds: a compound recognized for its part in the prevention of disease, known as lycopene is of particular importance and is in high demand from cosmetic, pharmaceutical, and food industries (Borguini and Torres, 2009) (Figure 5.3).

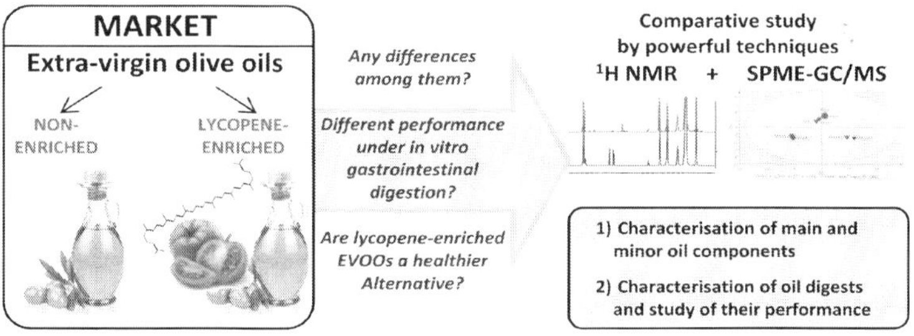


**Figure 5.3:** Extra-virgin oils are produced from tomato and olives.

*Source:https://www.sciencedirect.com/science/article/abs/pii/S0963 996920300120.*

Chemical synthesis or standardized tomato fruit extraction are the common sources of commercially accessible lycopene, but due to complicated extraction procedures, the product is normally expensive. Comparative to a pulp, the by-products have the peel fraction of tomato contains up to five-time additional lycopene, but the processing and storage of this material is an issue because of its susceptibility to microbial spoilage and high moisture level. The extractability of lycopene is extremely low by traditional food-grade organic solvents, such as ethyl acetate, ethanol, and hexane, at least under this situation that usually reserves the activity of the carotenoids *in vivo* (Zuorro and Lavecchia, 2010; Papaioannou and Karabelas, 2012). It was determined with pectinolytic enzyme preparation, an enzymatic pre-treatment of tomato peels, suggestively increased the level of lycopene recovery, together with the use of surfactant-assisted extraction. Moreover, for recovery of the lipophilic lycopene, the utilization of organic solvents, was evaded making the procedure of its extraction ecologically friendly, guaranteeing direct lycopene usage in cosmetics and food industries.

Olive (*Olea Europea*) oil is also a significant part of the Mediterranean diet, besides tomatoes connected with enhanced health status in consideration with reduced occurrence of coronary heart diseases. In an extensive variety of cosmetic products, it has been used as a vital ingredient like body and face creams, shampoos, soaps, lotions because of the existence of phenolic compounds and monounsaturated fatty acids. The aqueous phase of the olives predominantly contains these compounds and in epithelial and endothelial platelets, cells, neurons, neoplastic cells and cells of the immune system, they exert diverse biological activities (Cicerale et al., 2009).

The key phenolic compounds have powerful antioxidant activity both *in vitro* and *in vivo* including oleuropein and hydroxytyrosol and their derivatives, which provide extra-virgin olive oil its harsh taste. Regarding the oleuropein and its pharmacological activities, there are various studies including anti-atherogenic, anti-inflammatory, antiviral, anti-cancer, antioxidant, hypolipidemic, antimicrobial, and hypoglycemic effect. Oleuropein and hydroxytyrosol along with the antioxidant compounds exiting in the olives, only a low percentage dissolves in the lipidic fraction of the olives because they are highly soluble in water (Visioli and Galli, 2002; Omar, 2010).

Most of the polyphenols vanish in the Oil Vegetation Waters (OVW), during the olive oil extraction procedure, which shows the fraction of the olives which is soluble in water, which are detached from the oil. International Olive Council gives some figures that the consumption of olive oil enhanced 78% between 1990 and 2010 worldwide (International Olive Council, 2011). Proportionally, the increase in the olive oil has increased the olive mill by-products, since the olive oil is on average less than 20% of the olive fruit, comprising OVW. For human health, the OVW has valuable features, including anti-inflammatory activity, and it shows for various biotechnological applications, a huge source of useful compounds including Cosmetics (Manach et al., 1998; Bitler et al., 2005).

Effects of hydroxytyrosol have been proved in recent studies on cytokines, inflammatory mediators and chemokines, and anti-inflammatory constituents were identified in aqueous olive extracts (Richard et al.,

2011). High-Performance Liquid Chromatography (HPLC) is used for the fractioning of aqueous olive extracts and their effects on the production of inflammatory mediator's prostaglandin E2 (PGE2) and nitric oxide were investigated. It was determined that the production of PGE2 and NO was inhibited by the hydroxytyrosol, showing powerful anti-inflammatory activity and along with OVW decreases the secretion of chemokines (CXCL10/IP-10, CCL2/MCP-1) and cytokines (IL-1α, IL-1β, IL-6, IL-12, TNF-α). Furthermore, the valuable features of two fractions of OVW were investigated, OVW from *Olea Europea*, Carolea variety, in grouping with acacia fibers. The detoxifying effect, a strong anti-inflammatory effect, and antioxidant activity were monitored in on cultured skin cells, which suggested that fractions could be established as products for skincare, attained from OVW (Nasopoulou et al., 2014; Barbulova et al., 2015).

## 5.4. VALORIZATION OF BY-PRODUCTS DERIVING FROM PINEAPPLE AND GRAPEVINE

All the by-products and their derivatives, as already mentioned have some beneficial utilization, hence, concentration has been directed on how to attain greater value from them in consideration of efficiency. Grape mark extracts and pineapple rind and essential by-products are the examples of this type of valorization (Martín-Peláez et al., 2013; Yao et al., 2015).

A high quantity of by-products is produced from pineapple (*Ananas comosus*), which is the most processed tropical fruit. Being the rind and the core of the main ones, 25%–35% of the pineapple fruit is represented in these by-products (Reinhard and Rodriguez, 2009; WU et al., 2012). Compounds with high antioxidant activity are present in the pineapple rind and core, and concerning valorization in the cosmetics industry, they give them a huge potential. Presence of eight diverse carotenoids was determined by the use of ultra-high performance liquid chromatographic (UHPLC) technique (β-cryptoxanthin, zeaxanthin, lutein, β-carotene, violaxanthin, neoxanthin, α-carotene, and lycopene) and the result of Ultraviolet C (UV-C) radiations on these compounds and two vitamins (E and A) in the pineapple rind and core were assessed (Freitas et al., 2015).

One of the major sustainable sanitation methods of fruits is the ultra-violet radiation at a wavelength of 190–280 nm (UV-C), it is a relatively inexpensive and environmentally friendly method. It can promote the biosynthesis of vitamin C and carotenoids, can induce stress in the plant cells, and could provide additional benefits to the by-products (Gonzalez-Aguilar et al., 2010). Based on this, it was verified that L-ascorbic acid level in the whole fruit was lower compared with its level in the pineapple rind and it was also confirmed that L-ascorbic acid content in the rind is increased by the UV-C radiation. It was also confirmed that β-carotene was most abundant among the carotenoids, the carotenoid whose concentration also enhanced in the rind after exposure of UV-C radiation has the maximum pro-vitamin A activity (Figure 5.4).

**Figure 5.4:** Different beauty products manufactured from grapevine and pineapple.

*Source:https://i2.wp.com/ommorphiabeautybar.com/wp-content/uploads/2019/09/Caudalie-Skinare-2019-Launches-opener.jpg?fit=1149%2C1200.*

High nutritional value, rich of beneficial compounds is also given to another specie Grapevine having compounds such as carotenoids, tannins, polyphenols, and vitamins. Pomace or grape marc accounts for around 15% of the crushed grapes, the solid by-product which contains, skins, seeds, and pulp are leftover from the winemaking process (Caimari et al., 2013). Biotechnological processes utilize an industrial waste product represented as grape marc, uses are in the production of the Italian "Grappa," preparation of soil fertilizers, tartaric acid extraction and distillation of ethanol and purification of colorants (Carola et al., 2012). The grape marc is derived from white wine production which is obtained from Aglianico grapes (the most useful and the most esteemed variety of grapes in Campania region, Italy) and is richer in nutrients because no compound has yet been extracted by the winemaking process and it is unfermented. Furthermore, various types of lipophilic compounds are present in it, such as ellagic acid, resveratrol, carotenoids, lycopene, and at the time of pressing these are not dissolved in the water-soluble juice, but an oil-soluble (liposoluble) solvent is used for extraction and attained extract can be utilized as an active component for skincare Cosmetics (Oliveira et al., 2004).

A new cosmetic component was developed based on this assumption, used for making white wine Controlled Designation of Origin (DOC), Sant' Agata dei Goti, Benevento, Italy composed of a liposoluble fraction of the unfermented Aglianico marc. Another liposoluble extract attained from *Vitis vinifera* cells (Aglianico cultivar) was mixed with the liposoluble marc extract (LME), developed as sterile liquid cultures (liposoluble cell extract, LCE) and on cultured skin cells, the attained mixture was tested. The effect of the combination of the LCE and LME was studied on proteasome activity in human keratinocytes (HaCaT) and Human Dermal Fibroblasts (HDF), mixed in a 1:10 ratio. Elimination of damaged and abnormal proteins is the responsibility of "Proteasome" by the proteolytic system and its activity is inhibited in the stressed cells by the generation of aggregated and highly oxidized proteins in the cytoplasm (Bulteau et al., 2002). In enhancing the detoxifying capacity of the cells, the tested mixture of LCE and LME provided strong effects by up regulating the proteasome activity by 61% in keratinocytes and 44% in fibroblasts. In boosting the hydrating potential, the role of LCE/LME has also been proven by the investigators by measuring the expression level of HAS 2 and by measuring the expression level of Collagen I and Collagen III in stimulating new collagen production APQ3 (Aquaporin 3) and (Hyaluronan Synthase 2) genes.

## 5.5. COFFEE PROCESSING BY-PRODUCTS

The coffee processing industry is another source of the huge number of disposable by-products. In various countries, these coffee plants (*Coffea robusta* and *Coffea Arabica*) are being produced, particularly in those localized in South Asia, equatorial Latin America, Africa, and India. The elevated degree of processing knowledge is required in coffee dispensation and by-products such as coffee husk and pulp are produced, which have limited application as livestock feed, fertilizer, and compost (Murthy and Naidu, 2012). Contrary to this, quantities of unused green beans can be generated by the step of coffee berry selection and collection, because they cannot pass to the next processing step due to not sufficiently big and ripe or they have been mechanically damaged. These green beans are still considered a vital source of natural antioxidants, which have not yet been broken down by the roasting procedure (DeSA, 2013).

Hence, unroasted coffee beans were gathered, to have a valued ingredient, rich in bioactive anti-oxidant compounds, and using two natural elements carbon dioxide ($CO_2$) and pure water subjected to decaffeination, and extracted in ethanol. It was observed after the removal of ethanol that a jelly-like extract enhanced natural skin cell regeneration, reinforced the epidermal barrier and promoted an even skin tone for a radiant complexion. Furthermore, it also helped accelerated damaged skin repair, prevent water loss, and lighten the skin by inhibiting melanin synthesis by decreasing the inflammatory process (Flanagan et al., 2014; Bessada et al., 2018).

Moreover, it helped prevent water loss, lighten the skin by inhibiting melanin synthesis and accelerated damaged skin repair, by decreasing the inflammatory process (Flanagan et al., 2014; Bessada et al., 2018) (Figure 5.5).

**Figure 5.5:** Coffee-based beauty products.

*Source: http://whatalookerbeauty.blogspot.com/2017/10/my-top-10-coffee-inspired-beauty.html.*

## REFERENCES

1. Ayala-Zavala, J., Rosas-Domínguez, C., Vega-Vega, V., & González-Aguilar, G. A., (2010). Antioxidant enrichment and antimicrobial protection of fresh-cut fruits using their own by-products: Looking for integral exploitation. *J. Food Sci., 75*, R175–R181.
2. Baik, J. S., Kim, S. S., Lee, J. L., Oh, T. H., Kim, J. Y., Lee, N. H., & Hyun, C. G., (2008). Chemical composition and biological activities of essential oils isolated from Korean endemic *Citrus* species. *J. Microbiol. Biotechnol., 18*, 74–79.
3. Balkrishnan, R., McMichael, A. J., Hu, J. Y., Camacho, F. T., Shew, K. R., Bouloc, A., Rapp, S. R., & Feldman, S. R., (2006). Correlates of health-related quality of life in women with severe facial blemishes. *Int. J. Dermatol., 45*, 111–115.
4. Barbulova, A., Apone, F., & Colucci, G., (2014). Plant cell cultures as source of cosmetic active ingredients. *Cosmetics*, *2*, 94–104.
5. Barbulova, A., Colucci, G., & Apone, F., (2015). New trends in cosmetics: by-products of plant origin and their potential use as cosmetic active ingredients. *Cosmetics*, *2*(2), 82–92.
6. Bessada, S. M., C Alves, R., Oliveira, P. P., & Beatriz, M., (2018). Coffee silverskin: A review on potential cosmetic applications. *Cosmetics*, *5*(1), 5.
7. Bitler, C. M., Viale, T. M., Damaj, B., & Crea, R., (2005). Hydrolyzed olive vegetation water in mice has anti-inflammatory activity. *J. Nutr.135*, 1475–1479.
8. Borguini, R. G., & Torres, E. A., (2009). Tomatoes and tomato products as dietary sources of antioxidants. *Food Rev. Int., 25*, 313–325.
9. Bulteau, A. L., Moreau, M., Nizard, C., & Friguet, B., (2002). Impairment of proteasome function upon UVA- and UVB-irradiation of human keratinocytes. *Free Radic. Biol. Med., 32*, 1157–1170.
10. Caimari, A., Del, B. J. M., Crescenti, A., & Arola, L., (2013). Low doses of grape seed procyanidins reduce adiposity and improve the plasma lipid profile in hamsters. *Int. J. Obes., 37*, 576–583.
11. Carola, A., Tito, A., Bimonte, M., Mustilli, A., Cucchiara, M., Monoli, I., Hill, J., Apone, F., & Colucci, G., (2012). Liposoluble extracts of *Vitis vinifera* grape marc and cell cultures with synergistic anti-ageing effects. *HPC Today*, *7*, 42–46.
12. Cicerale, S., Conlan, X. A., Sinclair, A. J., & Keast, R. S., (2009).

Chemistry and health of olive oil phenolics. *Crit. Rev. Food Sci. Nutr., 49*, 218–236.

13. Costin, G. E., & Hearing, V. J., (2007). Human skin pigmentation: Melanocytes modulate skin color in response to stress. *FASEB J., 21*, 976–994.
14. DeSA, U. N., (2013). *World Population Prospects: The 2012 Revision* (Vol. 18, pp. 1–30). Population division of the department of economic and social affairs of the United Nations Secretariat, New York.
15. Djilas, S., Canadanovic-Brunet, J., & Cetkovic, G., (2009). By-products of fruits processing as a source of phytochemicals. *Chem. Ind. Chem. Eng. Q., 15*, 191–202.
16. Flanagan, J., Bily, A., Rolland, Y., & Roller, M., (2014). Lipolytic activity of Svetol®, a decaffeinated green coffee bean extract. *Phytotherapy Research*, *28*(6), 946–948.
17. Freitas, A., Moldao-Martins, M., Costa, H., Albuquerque, T., Valentea, A., & Sanches-Silvaa, A., (2015). Effect of UV-C radiation on bioactive compounds of pineapple (*Ananas. comosus* L. Merr.) by-products. *J. Sci. Food Agric., 95*, 44–52.
18. Goñi, I., & Hervert-Hernández, D., (2011). By-products *from Plant Foods are Sources of Dietary Fibre and Antioxidants, Phytochemicals—Bioactivities and Impact on Health* (Vol. 1, pp. 1–33). Available online: https://www.intechopen.com/books/phytochemicals-bioactivities-and-impact-on-health/by-products-from-plant-foods-are-sources-of-dietary-fibre-and-antioxidants
19. Gonzalez-Aguilar, G. A., Villa-Rodriguez, J. A., Ayala-Zavala, J. F., & Yahia, E. M., (2010). Improvement of the antioxidant status of tropical fruits as a secondary response to some postharvest treatments. *Trends Food. Sci. Technol., 21*, 475–482.
20. International Olive Council, (2011). *World Olive Figures* (Vol.1, pp. 1–30). Available online: http://www.internationaloliveoil.org/estaticos/view/131-world-olive-oil-figures (accessed on 30 May, 2020).
21. Kim, S. S., Lee, J. A., Kim, J. Y., Lee, N. H., & Hyun, C. G., (2008). Citrus peel wastes as functional materials for cosmeceuticals. *J. Appl. Biol. Chem., 51*, 7–12.
22. Kim, S., Kim, M. J., Choi, Y., Kim, B., Kim, K., Park, K., Park, S., et al., (2013). Down-regulation of tyrosinase, TRP-1, TRP-2 and MITF expressions by citrus press-cakes in murine B16 F10 melanoma. *Asian*

*Pac. J. Trop. Biomed., 3*, 617–622.

23. Kris-Etherton, P. M., Lefevre, M., Beecher, G. R., Gross, M. D., Keen, C. L., & Etherton, T. D., (2004). Bioactive compounds in nutrition and health-research methodologies for establishing biological function: The antioxidant and anti-inflammatory effects of flavonoids on atherosclerosis. *Annu. Rev. Nutr., 24*, 511–538.
24. Liu, Y. Q., Heying, E., & Tanumihardjo, S. A., (2012). History, global distribution, and nutritional importance of citrus fruits. *Compr. Rev. Food Sci. Food Saf., 11*, 530–545.
25. Manach, C., Morand, C., Crespy, V., Demigné, C., Texier, O., Régérat, F., & Rémésy, C., (1998). Quercetin is recovered in human plasma as conjugated derivatives which retain antioxidant properties. *FEBS Letters, 426*(3), 331–336.
26. Martín-Peláez, S., Covas, M. I., Fitó, M., Kušar, A., & Pravst, I., (2013). Health effects of olive oil polyphenols: Recent advances and possibilities for the use of health claims. *Molecular Nutrition and Food Research, 57*(5), 760–771.
27. Muracami, A., (2009). Chemoprevention with phytochemicals targeting inducible nitric oxide synthase. *Forum Nutr., 61*, 193–203.
28. Murthy, S., & Naidu, M., (2012). Sustainable management of coffee industry by-products and value addition: A review. *Resour. Conserv. Recycl., 66*, 45–58.
29. Nasopoulou, C., Karantonis, H. C., Detopoulou, M., Demopoulos, C. A., & Zabetakis, I., (2014). Exploiting the anti-inflammatory properties of olive (Olea europaea) in the sustainable production of functional food and neutraceuticals. *Phytochemistry Reviews, 13*(2), 445–458.
30. Oksman-Caldentey, K. M., & Inze, D., (2004). Plant cell factories in the post-genomic era: New ways to produce designer secondary metabolites. *Trends Plant Sci., 9*, 433–440.
31. Oliveira, C., Ferreira, A. C., Costa, P., Guerra, J., & Guedes, D. P. P., (2004). Effect of some viticultural parameters on the grape carotenoid profile. *J. Agric. Food Chem., 52*, 4178–4784.
32. Omar, S. H., (2010). Oleuropein in olive and its pharmacological effects. *Sci. Pharm.*, 133–154.
33. Papaioannou, E. H., & Karabelas, A. J., (2012). Lycopene recovery from tomato peel under mild conditions assisted by enzymatic pre-treatment and non-ionic surfactants. *Acta Biochimica Polonica, 59*(1),

1–40.

34. Papaioannou, E., & Karabelas, A., (2012). Lycopene recovery from tomato peel under mild conditions assisted by enzymatic pre-treatment and non-ionic surfactants. *Acta Biochim. Polonica, 59*, 71–74.
35. Pavithra, P. S., Sreevidya, N., & Verma, R. S., (2009). Antibacterial activity and chemical composition of essential oil of *Pamburus missionis*. *J. Ethnopharmacol., 124*, 151–153.
36. Reinhard, A., & Rodriguez, V. L., (2009). Industrial processing of pineapple-trends and perspectives. *Acta Hortic., 822*, 323–328.
37. Richard, N., Arnold, S., Hoeller, U., Kilpert, K., Wertz, K., & Schwager, J., (2011). Hydroxytyrosol is the major anti-inflammatory compound in aqueous olive extracts and impairs cytokine and chemokine production in macrophages. *Plant. Med., 77*, 1890–1897.
38. Sabio, E., Lozano, M., Montero, D. E. V., Mendes, R. L., Pereira, A. P., Palavra, A. F., & Coelho, J. A., (2003). Lycopene and β-carotene extraction from tomato processing waste using supercritical CO2. *Industrial and Engineering Chemistry Research*, *42*(25), 6641–6646.
39. Schieber, A., Stintzing, F., & Carle, R., (2001). By-products of plant food processing as a sour02ce of functional compounds-recent developments. *Trends Food Sci. Technol., 11*, 401–413.
40. Strati, I. F., & Oreopoulou, V., (2014). Recovery of carotenoids from tomato processing by-products: A review. *Food Research International*, *65*, 311–321.
41. Tuck, C., Perez, E., Horvath, I., Sheldon, R., & Poliakoff, M., (2012). Valorization of biomass: Deriving more value from waste. *Science*, *337*, 695–699.
42. Visioli, F., & Galli, C., (2002). Biological properties of olive oil phytochemicals. *Crit. Rev. Food Sci. Nutr., 42*, 209–221.
43. Wu, Z., Zhang, M., & Wang, S., (2012). Effects of high-pressure argon and nitrogen treatments on respiration, browning and antioxidant potential of minimally processed pineapples during shelf life. *J. Sci. Food Agric., 92*, 2250–2259.
44. Yang, E. J., Kim, S. S., Oh, T. H., Baik, J. S., Lee, N. H., & Hyun, C. G., (2009). Essential oil of citrus fruit waste attenuates LPS-induced nitric oxide production and inhibits the growth of skin pathogens. *Int. J. Agric. Biol., 11*, 791–794.
45. Yao, J., Zhang, B., Ge, C., Peng, S., & Fang, J., (2015). Xanthohumol,

a polyphenol chalcone present in hops, activating Nrf2 enzymes to confer protection against oxidative damage in PC12 cells. *Journal of Agricultural and Food Chemistry*, *63*(5), 1521–1531.

46. Yen, F. L., Wang, M. C., Liang, C., Ko, H. H., & Lee, C. W., (2012). Melanogenesis inhibitor(s) from *Phyla nodiflora* extract. *Evid. Based Complement. Alternat. Med.*, 1, 1–20, doi:10.1155/2012/867494.
47. Yilmaz, T., Kumcuoglu, S., & Tavman, S., (2016). Ultrasound assisted extraction of lycopene and β-carotene from tomato processing wastes. *Italian Journal of Food Science*, *29*(1), 1–30.
48. Zuorro, A., & Lavecchia, R., (2010). Mild enzymatic method for the extraction of lycopene from tomato paste. *Biotechnol. Biotechnol. Equip.*, *24*, 1854–1857.
49. Zuorro, A., Lavecchia, R., Medici, F., & Piga, L., (2014). Use of cell wall degrading enzymes for the production of high-quality functional products from tomato processing waste. *Chemical Engineering*, *38, 1–22*.

# 6

# *Use of Plant by-Products for Compost Manufacturing*

## CONTENTS

## 6.1. INTRODUCTION

Under thermophilic, aerobic conditions, the managed microbial degradation of organic material commonly defines the *Composting* (Haug, 1993). In this chapter, composting does not include anaerobic biodegradation and is considered an aerobic process. *Compost* is a stable, hummus-like product, the material result of the process, viable weed seeds and free of pathogenic organisms that are a valued agronomic soil amendment (Carlile, 2005) (Figure 6.1).

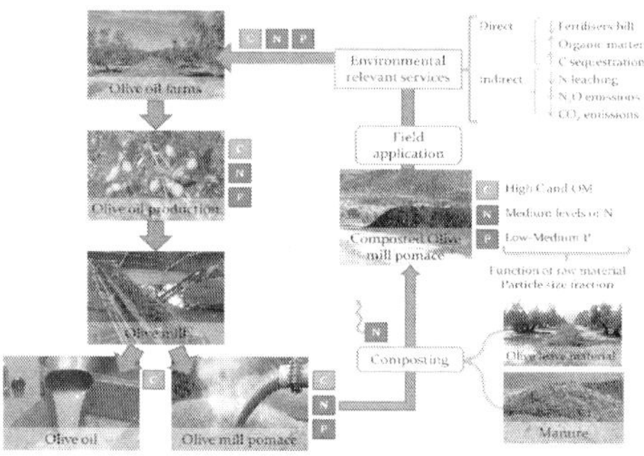


**Figure 6.1:** Composting process flowchart for olive oil farm and olive po-roducts.

*Source: https://www.intechopen.com/books/sustainable-development-authoritative-and-leading-edge-content-for-environmental-management/the-compost-of-olive-mill-pomace-from-a-waste-to-a-resource-environmental-benefits-of-its-applicatio.*

In many economic sectors, composting is getting popularity as an inexpensive technique for the valorization and processing of organic residuals. With extensively ranging levels of technical sophistication, it is practiced at numerous scales. In one case, the underlying concept is so easy and simple that can learn to manage and build a pile of compost in their home garden. In another scenario, the optimization of large-scale composting is a challenge because of various interacting chemical, physical, and biological processes at play (Zinati, 2005; Karadeniz and Kim, 2014). Hence, bridges are built by composting between municipal and agricultural, community, and individual, urban, and rural contexts. For civic pride for municipalities and broad public engagement, the reprocessing of organic residue into a valuable product can be a catalyst and it turns a profit for businesses and represents a positive public image and for individuals, it develops a sense of satisfaction and accomplishment (Stewart et al., 2000).A shift in terminology reflects the engagement in composting and the groundswell of public support, one that is also regularly happening in other features of waste management (Banna, 2014). "Resources" or "residuals" are the most common word used for organic by-products and words like "waste" or "garbage" are less often referred for these by-products. In this chapter, in recognition of the increasing concern for the desire to valorize organic by-products and responsible environmental stewardship, the term "organic residuals" is used, as part of the circular economy (Ellen MacArthur Foundation, 2017). On occasion in deference to the vast literature about waste management, the term "organic waste" is also is used (Camberato et al., 2006).

### 6.1.1. Agricultural Use of Compost

Perhaps since humans began to differentiate between unproductive and fertile, Composting has been accomplished for millennia, to improve and develop them for agriculture. Sir Albert Howard and Lady Gabrielle are the attributes for the leading scientific study of composting. During the early part of the twentieth century, Sir Albert was employed as an Imperial Economic Botanist in the erstwhile South Asian colonies of the British Empire (Howard, 1953). Agronomic benefits of composting were documented by this couple during this time, as part of the durable agricultural civilizations in areas of present-day India and China. Throughout the British Empire, composting was widely embraced following the endorsement of the Howards, (Haug, 1993) (Figure 6.2).

**Figure 6.2:** Agriculture composting site.

*Source: https://www.indiamart.com/team365-indore/compost-fertilizer.html*

Refinement and invention of the Haber-Bosch process in mainstream commercial agriculture, composting was subsequently eclipsed for fixing atmospheric nitrogen gas as ammonia (Smil, 2001; Nuhoff-Isakhanyan et al., 2016).

Ultimately synthetic nitrogen fertilizer was made convenient and inexpensive by the Haber-Bosch.

On the potential of abundant nitrogen fertilizer, during the Green Revolution, agricultural engineers and scientists capitalized in conjunction with improved irrigation, hybrids, and agricultural mechanization, to substantially raise the yields of the crop (Evenson and Gollin 2003).

The use of compost, regardless of many agronomic benefits, was largely limited to gardens and backyards for many decades. In commercial agriculture, nutrient management is still dominated by synthetic fertilizers; however, a renewed appreciation for the role of organic matter is produced, as mentioned below by the rising popularity of organic in maintaining good soil health, and other factors and use of compost at all scales in the agricultural are spurring renewed interest.

There is a determined need for impartial research of the drawbacks and benefits, in light of this resurgence of agricultural compost use (Raviv, 2010).

## 6.1.2. Municipal Waste Management

Compositing was identified as a technique of sanitizing and stabilizing by municipal waste management engineers, in the 1960s, from wastewater treatment plants.

Aquatic pollution was worsening by the increasing urban population by the release of raw sewage into surface waters during the twentieth century (Delleur, 2003).

In the wastewater treatment arrangement, a great amount of investment was done due to public outcry grew against this exercise. The novel infrastructure included sewage and storm water and instead of combined separated infrastructure and enhanced capacity of treatment of freshwater.

A huge amount of biosolids produced was managed by the resulting conundrum by the wastewater treatment plants.

Substitute to land filling the biosolids, Beltsville composting technique was established; instead, they are composted in aerated, static piles (discussed below) to generate a hygienic soil amendment (Haug, 1993) (Figure 6.3).

**Figure 6.3:** Composting as a viable strategy for solid waste management.

*Source: https://phys.org/news/2017-01-composting-viable-option-solid-sri.html.*

For the biosolids, a disposal option was offered by the Beltsville process, produced by the increasing number of sufficiently efficient wastewater treatment plants. However, for the compost, a market was required by the municipal engineers, an issue which many compost producers still face today. As popularly explained by the Howards, besides the agronomic advantages of compost and further deliberated below, there was and still, there is a public restlessness about agricultural utilization of compost that could comprise of pathogens and other harmful contaminants creates in portion from human ordure (Beecher et al., 2004).

Safe recycling of composted biosolids is a grave concern and research is ongoing on this issue and many authorities severely control their use and quality to safeguard public health (CCME 2005).

The suitability credo of the "throw-away" society and the inevitable increase of consumer capitalism, at the close of the twentieth century, began to grow into a disaster in solid municipal waste management, which endures in much of the world. It became very problematic to find new sites because the presently occurring landfills were getting capacity. More and more, the exercise of land filling garbage was considered as unsustainable and undesirable by the public. Hence substitute waste management exercises were encouraged for implementation, counting in the composting of the organic portion of municipal solid waste. Large and sufficiently sophisticated composting amenities began to construct by the regional and municipal governments' management authorities. Gardens, municipal parks, and street tree pits were utilized for the usage of compost produced by these facilities by the agricultural and horticultural sectors, and various other uses. At the time of writing, this trend remains in many areas, but in some areas, this development has thwarted by the emerged operational challenges (Martínez-Blanco et al., 2017; Wei et al., 2017).

### 6.1.3. Organic Agriculture, Agroecology, and Soil Conservation

The demand for organic" produce, by consumers, is a corresponding trend to the growing supply of compost, enhanced without the utilization of chemical pesticides or synthetic fertilizers. In developed economies, a noteworthy component of agricultural production has been accounted for by the organic agriculture, since its emergence in the mid-1900s and is a significant receiver of compost (Brzezina et al., 2017).

In agroecology, composting is also considered a significant tool, and to the industrialization of agriculture and the governance of multinational agribusiness benefits, it is a developing movement reactionary. Traditional agricultural practices are combined with modern science by the struggles of agroecology, minimizing the need for industrial inputs and emphasizing the natural cycling of nutrients by the development of sustainable agricultural landscapes (Gliessman, 2017). Hence, the local nutrient cycle can be enhanced by composting, in this context, by returning long-lasting, beneficial organic carbon to the soil (van Es, 2016).

### 6.1.4. Circular Economy

Turner and Pearce (1989) introduced a term *circular economy* that refers to an economic system in which components, products, and materials are deliberately managed and designed to preserve their maximum value at all times. Limited utilization of new resources is emphasized by the circular economy, where possible, the reuse of the products and reclamation and recycling of materials at the finish of their valuable life. Biological and technological are the two complementary resource cycles with composting as a conspicuous portion of the biological cycle in the circular economy (Ellen MacArthur Foundation, 2017).

Governments are now adopting the concept of the attendant aim of recycling organic residuals by composting and the idea of circular economy at the subnational, national, and supranational levels (e.g., Province of

Ontario, 2017; European Union, 2014). In the face of destabilizing trends, about the sustainability of our civilization, public concern is increasing such as unrestricted resource ingestion, the loss of biodiversity, global climate change and the eutrophication of the oceans. To develop likely mitigation strategies and these phenomena are considered by forcing the governments by their citizens. For responsible environmental stewardship and sustainable resource management policy a convenient conceptual framework is provided by the circular economy (Stahel, 2016). To operationalize such policies, programs are being rolled out, but, local governments have already implemented appropriate services and built organic waste management infrastructure, in various cases, for the reasons described previously (e.g., Gamble, 2005).

## 6.2. BENEFITS AND CHALLENGES

### 6.2.1. Agronomic

Micronutrients and macronutrients come from a good source of compost, although not as rich as synthetic fertilizers in readily bioavailable nitrogen (Hepperly et al., 2009). However, the nutrients present in it discharge slowly into the soil as the compost degrades and is relatively stable forms. Hence, rapid spikes of additional nutrients are prevented that could leach downward move into surface into groundwater water out of the root zone.

Along with the benefits of the systemic organic matter and nutrients recycling, other features of soil health are directly profited by compost. United States Department of Agriculture (USDA-NRCS 2012) defined soil health as, "the sustained capacity of the soil to function as a dynamic living ecosystem that withstands humans, plants, and animals." Soil health includes some important features like adequate depth to permit decent root growth; adequate but not unnecessary nutrient supply; good physical character or tilth; small populations of, pests, pathogens, and weeds; good drainage and water storage capacity; nonexistence of potential toxins and harmful chemicals; plentiful helpful organisms; and by potentially harmful influences such as hostile weather events, resistance to degradation (van Es, 2016). All these features are enhanced by compost. Moreover, it also increases soil volume. It contains intractable, humic organic compounds, which help to enhance the structure of soil, resilience, and water holding capacity (WHC). These compounds slowly release after absorbing nutrients ion due to the high cation exchange capacity of these compounds, also degrade the potentially toxic compounds by immobilizing their activity. Weed seeds and pathogens are inactivated by the microbial, thermophilic process and the compost is populated with the plentiful helpful microorganisms that boost the soil microbiome (Partanen et al., 2010).

### 6.2.2. Municipal

From composting, municipalities can profit in various means. By reducing the risk of environmental effects such as pollution of ground and surface water, the release of greenhouse gases and by averting organic materials from incineration or landfill, composting aligns with the social imperative of sustainability by its management option (Peigné and Girardin 2004). Materials that are considered as waste are valued by composting by changing into useful products for the municipality, to its citizens for its greening operations, and local horticultural and agricultural industries. Among residents and political capital, civic pride is generated by well-managed municipal composting for the administration (Ishii et al., 2000).

### 6.2.3. Circular Economy

Composting fits well into the heading of the circular economy, according to the developing point of view of holistic resource management and environmental stewardship (Ellen MacArthur Foundation, 2017). For the sanitization, recycling, and stabilization of organic material, composting is considered to be an efficient technology. The requirements of nonrenewable and energetically intensive nutrient supplements are reduced by the use of composts and hence the discharge of greenhouse gases is also reduced that would then be related to the manufacturing of those fertilizers. Compost is a stable amendment, beyond recycling and conserving of nutrients and organic matter, that raises the quantity of recalcitrant organic carbon in the soil (Wong and Fang, 2000).

### 6.2.4. Challenges

Composting is not appropriate for every scenario and is not a panacea. When evaluating composting as a selection for valorizing organic residuals, public safety and health challenges must be taken into account. To safeguard the public, most jurisdictions have guidelines or regulations in place by either governing the product quality or the process or both (Bernal et al., 2017). If the safety standards are not followed well, the possible survival of human pathogens is a principle concern in compost. Particularly in industrial or municipal compost, potentially toxic materials could be present such as trace metals or could stubborn organic pollutants such as hormones, antibiotics, and endocrine disruptors. Materials that degrade very slowly or even do not degrade, in the environment, and sharp foreign (inorganic) objects are some other possible hazardous contaminants in the environment (MacGregor et al., 1981).

The health of crops can also be threatened by poor-quality compost (Bernal et al., 2017). Some of the similar or same hazards as those described concerning humans can also harm the plants. Weed seeds or plant pathogens might endure an inappropriately managed composting process, and organic and inorganic toxicants can be a predator for plants. Extra application of compost, can deplete oxygen, particularly if it is not fully mature, to decompose in the soil. Over time, any non-degradable fragments could accumulate in the soil that it contains, which might present a threat to wildlife and would be esthetically unpleasing (Finstein et al., 1986).

## 6.3. COMPOSTING PROCESS

The conditions for healthy microbial activity are related to the necessities for effective composting and, unusually, correspond to four elements of Western alchemy: fire (heat), earth (feedstock), air, and water. These requirements make the management of the process challenging and complex and are interrelated (Figure 6.4).

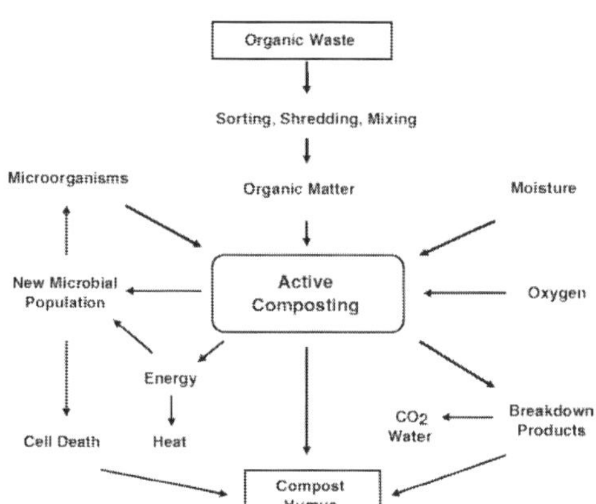


**Figure 6.4:** Amendment of soil using composting process for lignocelluloses waste material.

*Source: https://link.springer.com/chapter/10.1007/978-3-642-19769-7_6.*

### 6.3.1. Feedstock

Extra ingredients will be needed if the compost substrate does not contain a satisfactory amount of nutrients for effective co-composting. Significantly, nitrogen, and carbon must be provided by the feedstock in a mass ratio (N: C ratio) of about 1: 25 T (Haug, 1993).

This ratio is routinely assessed based on total nitrogen and carbon, but there is no need for the full amounts to be bioavailable readily; for example, materials such as hemicellulose, cellulose, and lignin are intractable (Anwar et al., 2017). On one hand, feedstock which comprises of woody materials, to sustain a suitable nutrient balance, the C:N ratio might be adjusted upward. On the other hand, green plant material normally has a reduced C:N ratio, but both these elements are bioavailable. In most organic materials, micronutrients are typically existing in satisfactory quantities but might need to be supplemented in exceptional cases (Sánchez et al., 2017).

For smooth drainage and airflow, the suitable physical structure must be provided by the feedstock, so during the composting process, the substrate does not compact exceptionally (Haug, 1993).

The amorphous substrate can be balanced by the addition of an appropriate bulking agent such as woody residuals. Shredded tires which are a synthetic material, can also be used and for reuse, at the end of the process, it can be screened out easily.

### 6.3.2. Aeration

In the compost bed, it is necessary to sustain an aerobic environment, related to the physical structure of the substrate. Aerobic conditions are ensured by an oxygen concentration of 5% or more (Haug, 1993).

For volatile metabolites and carbon dioxide removal, suitable airflow is necessary, also it is required to remove heat or excess moisture released during the metabolism of the substrate. Convection through the porous substrate is the basis of the maintenance needed for airflow, probably helped by the agitation of the compost or periodic turning or by mechanical ventilation in some systems.

However, compost can be cooled or dried by the excessive aeration, hence, slowing the process. Nitrogen can be stripped in the form of ammonia a nuisance gas from the substrate by the high aeration rates, and its loss reduces the value of fertilizer of the finished product.

### 6.3.3. Temperature

Composting is a thermophilic process, as clearly defined in this chapter. More than 45°C, thermophilic temperatures are usually attained in a well-managed process after a short variation phase of a few days, depending on the composition of compost, the mass of the compost pile and the form of the facility (Haug, 1993). During a "high-rate" period of numerous weeks, such temperatures will be maintained by the appreciable mass by the exothermic reactions of a well-managed composting system. As certainly degradable compounds are drained, a gradual reduction in the temperature will be followed. Then composting activity will be followed by a mesophilic and prolonged curing time during which ultimately the substrate stabilizes (Rosenfeld et al., 2007).

Thermophilic temperatures will terminate pathogenic organisms because this temperature is considerably above human body temperature, after a contact time dependent on a particular organism, the substrate and the temperature (Bernal et al., 2017). At thermophilic temperatures, weed seeds or insect larvae are also inactivated or killed, which is a significant feature for the product if it is to be used for the horticultural or agricultural sector. The compost should be insulated or turned in a way that confirms all the substrate reaches thermophilic temperatures and so there are no exterior layers or pockets that are ineffectively heated.

The composting process tends to slow down by the temperatures above 70°C, as the proteins in the cytoplasm of the microbes begin to denature and their metabolism is impaired. By increased aeration or by turning the compost, the extreme heat can be relieved. However, Hyperthermophilic composting processes that are modified to temperatures above 80°C which integrate microorganisms insulated from thermal springs, have been established (Kanazawa et al., 2006). Hyperthermophilic composting can practice feedstock more quickly than traditional thermophilic composting since at higher temperatures, the biological processes incline to function quicker and hence, a smaller footprint is required to process equal quantities of the substrate (Rosenfeld and Suffet, 2004).

### 6.3.4. Moisture

The moisture content of the substrate should be given careful attention for effective composting. Of the total wet mass, approximately 60% is the optimum moisture content for composting (Haug, 1993). On one side of the story, the extra moisture can hinder aeration and in the substrate, the pore space can be blocked. The process will become anaerobic and the microbial population will lack oxygen, causing cooler, much slower and stinking process that does not harvest suitable compost. Furthermore, from overly moist compost, water is likely to drain as leachate, producing challenges associated with storage, collection, and recycling or treatment.

On the other hand, due to a lack of moisture, inadequate water content also decreases the microbial activity for transport and dissolution of nutrients. Cracks can also be developed by the drying process and preferential airflow through the cracks in the substrate, which by producing a vicious cycle, exaggerates local drying. Uneven composting and inadequate aeration are caused by the enhanced airflow through the dry compost in other portions of the compost bed (Gilley and Van Durme, 2002). In these situations, from the air inlet, through the compost, a drying front can advance. A low moisture-holding capacity has been observed in the relatively cool ambient air. Moisture evaporates and increases in air's moisture-holding capacity is observed, when heated by the compost from the surrounding substrate. Hence, as the compost dries and cools, the microbial activity is depressed as a result. Then, incoming air before being humidified and warmed travels ever further into the compost bed. In a compost bed, heterogeneous moisture content can be remedied by mixing. Drying can be prevented in its first place by humidification or irrigation of the compost.

### 6.3.5. Management and Quality Assurance

The importance of proficient management of the composting process is equal to that of the physical process parameters as described above, or better considered as the prerequisite to them (Bernal et al., 2017). Experience and education both are required in management expertise, because of the distinctive features of composting operation. Due to the variable and extremely heterogeneous nature of the composting substrates, this is truer of composting than of many other processes. Adequate observation is required for the effective control of any process, especially so far for the composting process.

For quality assurance and process control or constant quality development, record keeping and monitoring are indispensable. Depending on the authority large-scale composting will perhaps be subject to obligatory reporting and permitting requirements (Walker et al., 2006).

Optimization of valorization and transformation of organic residuals is one aspect of compost management. Understanding of the obtainable feedstock is required in terms of physical structure, moisture, and nutrient content, and degradability as decried above. By season or by source, these features may differ, or in some cases, even by week or time of day. Stockpile and mixing of the feedstock may be required to compensate for the difference, or co-compost or mix various feedstock's, bulking agents or amendments in variable extents. Other factors such as weather events or seasonal climatic changes, apart from changes in feedstock, can affect the demand compensating management approaches and the composting process.

The assurance of the safety and health of the public and especially workers and risk is another aspect of management. Exposure to that hazard or even presence of hazard area the causes of risk. Therefore, either by prevention of exposure or removal of the hazard is involved in the mitigation of risk. On-site production of toxic gases such as sulfide and hydrogen, the existence of pathogens and material properties of the compost or feedstock, or the machinery or structures used on the site can be related to the risks to workers (Brown, 2016). By eliminating such hazards, occupational health and safety guidelines are meant to guard workers so far as possible. To stop exposure of the workers to the hazard, personal protective equipment made accessible to them and suitable measures should be put in place, where removal of a hazard is impossible.Elimination of the hazard is the preferable choice when there is a potential risk to the public as with worker safety. To ensure safe compost, the procedure must be familiar with any applicable regulations and in many jurisdictions, the guidelines exist for process control (CCME 2005). By maintaining a temperature above a specific level for a particular period, then the goal elimination of human pathogens of such guidelines can be achieved. To ensure satisfactory heat treatment of all the material, mixing or turning of the might also be recommended. Giddiness for the quality of the end product also exists along with the characterization of the process. The presence maximum numbers of indicators organisms or potentially pathogenic organisms are usually specified by the product quality guidelines, the pathogens may include fecal coli form bacteria, etc.

The size and number of foreign particles might also be specified by these guidelines in a sample of finished compost; the concentrations of potentially toxic metals; the existence of organic contaminants, and the stability or maturity of the compost. Specific testing protocols are usually referred to or are included in these guidelines for each appropriate measure. Ultimately, there might be restrictions on how and where that product can be used, where a hazard cannot be eliminated from a product

## 6.4. COMPOST FACILITIES

Several various styles of composting are present, each of which is suitable for diverse contexts and scales (Haug, 2013; Haug, 1993). There are various factors on which the type of the processes is selected which include, surroundings, market conditions, location, regulatory, and social context, feedstock, and the accessibility of labor. According to the kind of process, the cost of maintenance, operation, and capital varies and so hence the feasibility of the process is also indicated by the finances. Careful consideration must be given to the maintenance, durability, ultimate end-of-life replacement and repair, in the technologically sophisticated and capital-intensive operations. Metallic structures and machinery are susceptible to corrosion, for example, which can be enhanced from active composting of nitrogen-rich feedstock by exposure to ammonia emissions (Kashmanian and Spencer, 1993) (Figure 6.5).

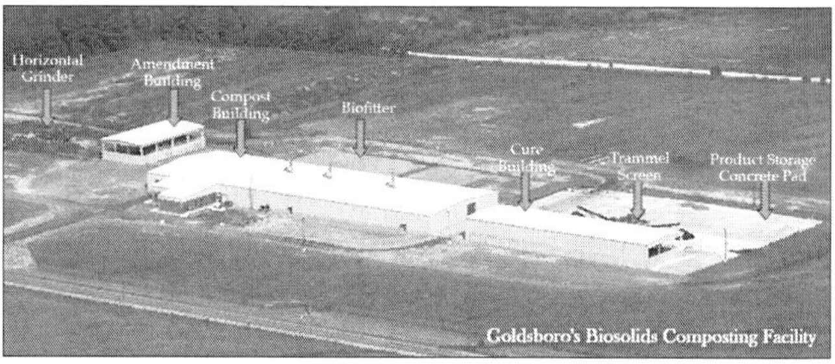


**Figure 6.5**: Typical structure of a composting facility (Goldsboro biosolid composting facility).

*Source: https://compost-turner.net/composting-technologies/in-vessel-composter-and-windrow-turner.html.*

### 6.4.1. Windrow

Turned windrow is the simplest type of large-scale composting. To ensure suitable initial conditions, the compost ingredients are mixed, as mentioned previously. A windrow is then formed from the mixed feedstock which is narrow piles, long with a trapezoidal or triangular cross-section, normally, several meters across and less than 2 m high. For oxygen concentration, temperature, and moisture content, the windrows are monitored. By the use of machineries such as specialized equipment, such as a windrow turner or a tractor with a front-end loader, the windrows are turned frequently. During the thermophilic phase, high-rate to promote aeration, homogenize the substrate and control temperature, the windrows should be turned about once a week. To sustain an appropriate moisture content, it might be essential to add water during turning. As the substrate cools and composting progress, turning can be less frequent. Ultimately, the compost cures for as long as numerous months, once the temperature has neared ambient and might then continue stockpiled until used or sold (Wolsink, 1994).

### 6.4.2. Aerated Static Pile

In broad, short piles the mixed compost typically is shaped, in an aerated static pile system. When the piles are built, a technique is used to pull or blow the air through the compost bed, in the bottom of the pile through perforated pipes, an aerated floor or a porous sublayer. As mentioned previously, an aerated static pile assembled on a layer of the porous substrate is contained in the Beltsville system. Air is drawn through perforated pipes laid in the porous substrate by the mechanical blower, and the air is exhausted through a pile of wet woodchips or similar materials that is used as a biofilter to eliminate odorous gases. The compost pile is not mixed as the name suggests, but instead throughout the process, remains static. Active compost in the pile is insulated and covered by a layer of finished compost to ensure that the thermophilic temperatures are reached to the whole mass of the feedstock. To decrease odors produced from the pile, the layer of finished compost also acts as a biofilter (Kaplan et al., 2004).

Conflicting to the Beltsville system, which functions under negative pressure, commercial systems are existing, in which a static aerated pile functions under positive pressure. Synthetic textile semi permeable covers the pile, restricting the flow of moist air and allowing its passage to dispense it evenly throughout the pile. As compared to the Beltsville system, this kind of system is advantageous as less energy and smaller blower are required to dispense airflow throughout the compost bed evenly. The replacement costs, as well as initial cost of the textile cover, are considerable. However, to facilitate the deployment of the cover, the apparatus is also needed over the pile and its following stowage at the end of the active composting phase.

### 6.4.3. Channel Composting

Composting is more capital-intensive in bunkers or channels than a static pile or windrow systems, but due to the constrained cross-section of the compost pile by the wall the footprint for such systems is smaller and is therefore rather than trapezoidal it is rectangular (Diaz et al., 2011). Windrows are shallower than channel compost bed, for example, bunkers, and channels offer suitable control over process conditions since aeration is assisted mechanically. As with a static pile, satisfactory aeration of the compost must be provided by the facility, normally through an aerated floor on which machinery drives for unloading and loading. Batch operations are performed as all-in and all-out in such systems. As the compost matures, the conditions for the process can be adjusted for the entire batch, for the entire batch needing changing rates of irrigation and aeration (Cheung, 2015).

### 6.4.4. In-vessel Composting

Compared to open systems, composting vessels are more capital-intensive (Diaz et al., 2011). However, greater control is offered over the process and hence, optimization can be done for better and faster processing, ensuing a more compact footprint.

The containment of odors and their smaller relative size makes them more appropriate for the more public locations. A wide variety of configurations are offered by the composting vessels, but a horizontal cylinder is a common type with a slight incline.

The entire cylinder or an internal auger rotates irregularly around its longitudinal axis and the vessel, at the elevated end is loaded through an access door, and gradually moving the compost toward the unloading port down the incline. Aeration and moisture can be precisely organized because the vessel is closed, the entire mass of the compost is systematically heated and the internal temperature is uniform. Any unprocessed aggregates can be retained inside the vessel that reaches the outlet of the container by a screen or else recycled to the inlet (Haug 199; Gamble, 2005).

## 6.5. SITING CONSIDERATIONS

It is wise to take into account the larger context in which the facility operates when planning a composting facility (de Bertoldi-Schnappinger, 2011).

The viability of the whole project can be put into risk by the primary attention on the narrow technical considerations of the actual process because larger concerns might be neglected by the planners that govern the long-term sustainability of the operation.

### 6.5.1. Regulations

Perhaps the most important aspect is familiarization with the regulatory environment, apart from developing a sound business plan of starting a new composting operation (Bernal et al., 2017; CCME 2005). On one side of the story, there may be substantial permitting costs and critical restrictions and on the other side, published guidelines and published guidelines can be priceless sources of information and advice about the social environment and business, management strategies and appropriate technologies. National, regional (departmental, provincial, state) and local legislation could all be appropriate, and therefore, all levels of government should be referred. A professional consultant is indispensable, therefore, when planning an operation of an appreciable scale to effectively navigate regulatory complexities.

### 6.5.2. Space

When consideration is shifted to the selection of the location from the regulatory environment, various physical reflections come into play. Again, perhaps most important concerns are those of the largest scale for the long-term success of the project and therefore, are likely to impinge on neighboring businesses or residents.

Any necessities for co-composting constituents and quantity of feedstock to be composted, the overall space requirements will be dictated by the curing, bulking agents and stockpiling of the complete product. A primary driver in site selection will be the essential footprint for the complete operation and, municipal zoning restrictions could come into play, depending on the scale of the operation (Meunchang et al., 2005).

### 6.5.3. Traffic

To bring feedstock and other needed materials to the site and to move the end product, the amount of traffic necessary is associated with the scale of the operation. Concern about the safety of pedestrians and drivers is raised by the amplified traffic.

Drivers might feel unease by the regular passage of possibly heavy vehicles that could cause congestion and along with that rise in requirements of road maintenance. Dust, increased traffic noise and light pollution could surely be harmful in a commercial or residential neighborhood.

### 6.5.4. Odor

A frequently overlooked consideration is odor control and reduction and often is a cause of unwell planned composting facilities. The process and feedstock are potential sources of nuisance odors, although end product compost has an inoffensive odor. Therefore, odor control must be incorporated and is imperative to the earliest design stages process and of the site location (Bidlingmaier and Müsken, 2011). The potential for odors and other nuisances will be greatly reduced by the onsite conscientious management and cleanliness and also improve overall productivity and safety. When planning the location of operations, the local topography, and prevailing winds must also be taken into account on the site. Remote areas should be selected away from commercial buildings and commercial buildings for operations like mixing and grinding of feedstock and other constituents, the stockpiling, etc. From roadways and public views, to screen potentially odorous operations it is wise to use fences or trees. Such screens will also decrease odor dispersion and serve as windbreaks. It might even be essential that odorous operations should be performed inside a closed structure from which the exhaust air can be treated in the container house before discharge.

Compared with the treatment of putrescible feedstock, the composting process itself, if well managed, will probably be less odorous. Within the compost pile, there is potential for anaerobic pockets that produce malodors, however, even the best operations are never ideal.

The apparatus and kind of composting process, therefore, should be selected cautiously to ensure that odors are not an issue either off or onsite. The risk of nuisance odors could be elevated for instance by closeness to a residential neighborhood enough to permit investment in a negative-pressure system such as an enclosed process such as an agitated bed, bunker, channel, aerated static piles, or even in-vessel composting. The possibility of complete containment is offered by the in-vessel or enclosed systems to treat all process air (Wu et al., 2011).

Lastly, Implementation of in-house, a proactive program is a significant part of odor management (Laor et al., 2014).

By installing a highly visible, well-advertised, and effortlessly accessible reporting system, the active involvement of the public should be ensured. To deal with any or incidents, an effective and rapid response protocol is needed.

### 6.5.5. Pests

To safeguard the safety and health of the workers, preserve good community relations and guarantee public hygiene and health, insect, and animal controls are important (Haug, 1993). Rodent's controls should be considered, as they are haggard to nesting sites and ready food sources and can carry disease in the warm compost. In untidy yards, ill-managed compost beds and stockpiles of putrescible feedstock, the insects can be problematic. In open facilities, birds also pose probable difficulties, particularly those situated near airports, because they interfered with the traffic. Eliminating or reducing stockpiles of feedstock, maintaining a tidy and clean yard and enclosing or covering operations that might appeal to animals can decrease the occurrence of harms. Along with the ability and readiness to respond quickly, constant vigilance is necessary when problems arise (Dao et al., 2001).

### 6.5.6. Water and Leachate

An important characteristic of site design is the management of leachate and surface water that is normally strongly regulated (Ontario Ministry of the Environment, 2012). If the site is situated downslope from poorly managed or poorly drained property, then consideration of run-on water is potentially significant. Drainage or diversion infrastructure construction is advisable such as swales or swales to prevent erosion of any windrows or windrows and inundation at the upper edge of the property for instance.

## 6.6. PRODUCTS FROM COMPOSTING

### 6.6.1. Compost

Composting can produce various products, the apparent one being the compost itself. With many agronomic profits, compost is a helpful soil amendment, as described previously, and for bulk sales of compost, there may be local markets for horticultural and agricultural use (Eggerth et al., 2011). In other substitute markets, there also exist some applications such as construction to endorse the establishment of vegetation and to alleviate exposed soil surfaces on construction sites. In other applications, to develop or improve the growth of plants, compost can also be beneficial, as the microbial activity is stimulated in the soil and on landfill covers or for mine site reclamation, as in soil decontamination or brownfield reclamation (Kumar et al., 2010).

### 6.6.2. Compost Tea

Water extract of finished compost is known as compost tea. As the compost soaks after bubbling air through water, it can be prepared aerobically or in sealed containers anaerobically (Scheuerell and Mahaffee, 2006). As a foliar spray, compost tea can be used directly to plants or the soil. Suspended particulates, soluble organic compounds, and microbes from the compost are present in it, the precise nature of these may vary based on the mode of preparation and on compost itself. Some claims overpower plant pathogens because certainly there are some nutrients in compost tea, it can also enhance the soil microbiota and promotes growth. About these supposed benefits, there is much anecdotal literature, however, currently, the burgeoning scientific literature about them is inconclusive (Sánchez-Monedero et al., 2010).

### 6.6.3. Heat

The capture of heat for use in nearby facilities is a developing trend in some larger composting processes (Walther et al., 2017). Latent and sensible heat is removed by the airflow during the active phase of composting and, to improve the process, the removal of surplus heat is essential. As the process temperature during composting is not more than 70°C, the subsequent thermal energy is low-grade. However, the building can be warmed by the process air, for example, or thermal energy could be captured from the compost bed or the process air by a heat exchanger and then warm water or air. Hence, the utilization of more costly energy sources can be completely or partly replaced by compost heat such as gas or electricity.

### 6.6.4. Gases

"Short-cycle" carbon dioxide or biogenic can be released by the composting process, which has benefits in some contexts. For example, carbon dioxide enrichment is used commonly by greenhouse operators to promote the growth of the plant. Process air is often saturated with humidity from composting can comprise of some trace gases, such as ammonia and methane, so depending on the application, dehumidification or washing might be needed before use (Droussi et al., 2009).

## REFERENCES

1. Anwar, Z., Irshad, M., Bilal, M., et al., (2017). Changes in availability of plant nutrients during composting of cow manure with poplar leaf litter. *Compost Science and Utilization, 25(*4), 242–250. https://doi.org/10.1080/1065657X.2017.1300547 (accessed on 30 May, 2020).
2. Banna, F., (2014). Organic waste as a valuable resource: A call for action. In: *Sustainable Cities* (Vol. 1, pp. 1–33). Washington: The World Bank Group. https://blogs.worldbank.org/sustainablecities/organic-waste-valuable-resource-call-action (accessed on 30 May, 2020).
3. Beecher, N., Connell, B., Epstein, E., et al. (2004). *Public Perception of Biosolids Recycling: Developing Public Participation and Earning Trust* (Vol. 1, pp. 1–22). Alexandria, VA: Water Environment Research Foundation. https://static1.squarespace.com/static/54806478e4b0dc44e1698e88/t/548a16f8e4b0233b56ad46d2/1418335992433/BeecherEtAl-PublicPerceptions-WERF04.pdf (accessed on 30 May, 2020).
4. Bidlingmaier, W., & Müsken, J., (2011). Odor emissions from composting plants. In: *Compost Science and Technology* (Vol. 1(11), pp. 215–324). Elsevier Science.
5. Brown, N. J., (2016). *Composting Safety and Health* (Vol. 1, pp. 1–30). Workplace Health and Safety Program. Ithaca, NY: Cornell University.
6. Brzezina, N., Biely, K., Helfgott, A., et al., (2017). Development of organic farming in Europe at the crossroads: Looking for the way forward through system archetypes lenses. *Sustainability, 9*, 810–821. https://doi.org/10.3390/su9050821 (accessed on 30 May, 2020).
7. Camberato, J. J., Gagnon, B., Angers, D. A., Chantigny, M. H., & Pan, W. L., (2006). Pulp and paper mill by-products as soil amendments and plant nutrient sources. *Canadian Journal of Soil Science*, *86*(4), 641–653.
8. Carlile, W. R., (2005). The use of composted materials in growing media. In: *International Symposium on Growing Media* (Vol. 779, pp. 321–328).
9. CCME, (2005). *Guidelines for Compost Quality* (Vol. 1, pp. 1–30). Winnipeg: Canadian Council of Ministers of the Environment. https://www.ccme.ca/files/Resources/waste/organics/compostgdlns_1340_e.pdf (accessed on 30 May, 2020).
10. Cheung, L., (2015*). Feasibility and Cost Analysis of Composting*

*Campus Dining Food Waste in Cal Poly's Compost Facility* (Vol. 1, pp. 1–22).

11. Dao, T. H., Sikora, L. J., Hamasaki, A., & Chaney, R. L., (2001). Manure phosphorus extractability as affected by aluminum-and iron by-products and aerobic composting. *Journal of Environmental Quality*, *30*(5), 1693–1698.
12. De Bertoldi-Schnappinger, U., (2011). Design of composting plants. In: *Compost Science and Technology*, (Vol. 1(6), pp. 89–117). Elsevier Science.
13. Delleur, J. W., (2003). The evolution of urban hydrology: Past, present, and future. *Journal of Hydraulic Engineering, 129*, 563–573.
14. Diaz, L. F., Savage, G. M., Eggerth, L. L., & Chiumenti, A., (2011). Systems used in composting. In: *Compost Science and Technology* (Vol. 1(5), pp. 67–87). Elsevier Science.
15. Droussi, Z., D'Orazio, V., Hafidi, M., & Ouatmane, A., (2009). Elemental and spectroscopic characterization of humic-acid-like compounds during composting of olive mill by-products. *Journal of Hazardous Materials*, *163*(2–3), 1289–1297.
16. Eggerth, L. L., Diaz, L. F., Chang, M. T. F., & Iseppi, L., (2011). Marketing of Composts. In: *Compost Science and Technology* (Vol. 1(12), pp. 325–357). Elsevier Science.
17. Ellen MacArthur Foundation, (2017). *Urban Biocycles*. Cowes, UK. https://www.ellenmacarthurfoundation.org/publications/urban-biocyles.
18. European Union, (2014). *End-of-Waste Criteria for Biodegradable Waste Subjected to Biological Treatment Compost and Digestate, Technical Proposals* (Vol.1, pp. 1–30). Joint Research Centre Scientific and Policy Reports. https://ec.europa.eu/jrc/en/publication/eur-scientific-and-technical-research-reports/end-waste-criteria-biodegradable-waste-subjected-biological-treatment-compost-digestate
19. Evenson, R. E., & Gollin, D., (2003). Assessing the impact of the green revolution 1960 to 2000. *Science, 300*, 758–762. https://doi.org/10.1126/science.1078710 (accessed on 30 May, 2020).
20. Finstein, M. S., Miller, F. C., & Strom, P. F., (1986). Monitoring and evaluating composting process performance. *Journal (Water Pollution Control Federation), 1*, 272–278.

21. Gamble, S., (2005). Five years of composting in Edmonton, Alberta. *Biocycle, 46(*10), 30.
22. Gilley, A. D., & Van, D. G. P., (2002). Biofilters: "low" maintenance does not mean "no" maintenance. In: *Proceedings of the Water Environment Federation, Odors and Toxic Air Emissions 2002*, (pp. 727–737). Water Environment Federation https://www.researchgate.net/scientific-contributions/2003147941_Alicia_D_Gilley.
23. Gliessman, S., (2017). A brief history of agroecology in Spain and Latin America. *Journal Agroecology and Sustainable Food Systems, 41*, 229–230. https://doi.org/10.1080/21683565.2017.1292390 (accessed on 30 May, 2020).
24. Haug, R. T., (1993). *The Practical Handbook of Compost Engineering*. Lewis/CRC.
25. Haug, R. T., (2013). Compost plant design and process management. In: *The Science of Composting* (Vol. 1, pp. 60–70). Springer Science & Business Media.
26. Hepperly, P., Lotter, D., Ziegler, U. C., et al., (2009). Compost, manure and synthetic fertilizer influences crop yields, soil properties, nitrate leaching and crop nutrient content. *Compost Science and Utilization, 17(*2), 117–126. https://doi.org/10.1080/1065657X.2009.10702410 (accessed on 30 May, 2020).
27. Howard, L. E., (1953). *Sir Albert Howard in India* (Vol. 1, pp. 1–33). London: Faber and Faber.
28. Ishii, K., Fukui, M., & Takii, S., (2000). Microbial succession during a composting process as evaluated by denaturing gradient gel electrophoresis analysis. *Journal of Applied Microbiology*, *89*(5), 768–777.
29. Kanazawa, S., Ishikawa, Y., Tomita-Yokotani, K., et al. (2006). Space agriculture for habitation on Mars with hyper-thermophilic aerobic composting bacteria. *Advances in Space Research, 41*, 696–700.
30. Kaplan, P. Ö., Barlaz, M. A., & Ranjithan, S. R., (2004). A Procedure for Life-Cycle-Based Solid Waste Management with Consideration of Uncertainty. *Journal of Industrial Ecology*, *8*(4), 155–172.
31. Karadeniz, F., & Kim, S. K., (2014). Trends in the use of seafood processing by-products in Europe. In: *Seafood Processing* By-Products (Vol. 1, pp. 11–19). Springer, New York, NY.
32. Kashmanian, R. M., & Spencer, R. L., (1993). Cost considerations

of municipal solid waste compost: Production versus market price. *Compost science and Utilization*, *1*(1), 20–37.

33. Kumar, R., Verma, D., Singh, B. L., & Kumar, U., (2010). Composting of sugar-cane waste by-products through treatment with microorganisms and subsequent vermin composting. *Bioresource Technology*, *101*(17), 6707–6711.
34. Laor, Y., Parker, D., & Pagé, T., (2014). Measurement, prediction, and monitoring of odors in the environment: A critical review. *Reviews in Chemical Engineering, 30(*2), 139–166. https://doi.org/10.1515/revce-2013-0026 (accessed on 30 May, 2020).
35. MacGregor, S. T., Miller, F. C., Psarianos, K. M., & Finstein, M. S., (1981). Composting process control based on interaction between microbial heat output and temperature. *Appl. Environ. Microbiol.*, *41*(6), 1321–1330.
36. Meunchang, S., Panichsakpatana, S., & Weaver, R. W., (2005). Co-composting of filter cake and bagasse; by-products from a sugar mill. *Bioresource Technology*, *96*(4), 437–442.
37. Minasny, B., Malone, B. P., McBratney, A. B., et al. (2017). Soil carbon 4 per mille. *Geoderma, 292*, 59–86. https://doi.org/10.1016/j.geoderma.2017.01.002 (accessed on 30 May, 2020).
38. Nuhoff-Isakhanyan, G., Wubben, E. F., & Omta, S. W. F., (2016). Sustainability benefits and challenges of inter-organizational collaboration in Bio-Based business: A systematic literature review. *Sustainability*, *8*(4), 307.
39. Ontario Ministry of the Environment, (2012). *Guideline for the Production of Compost in Ontario* (Vol. 1, pp. 1–33). Toronto, ON: Ontario Ministry of the Environment, Waste Management Policy Branch. http://www.ontario.ca/environment (accessed on 30 May, 2020).
40. Partanen, P., Hultman, J., Paulin, L., Auvinen, P., & Romantschuk, M., (2010). Bacterial diversity at different stages of the composting process. *BMC Microbiology*, *10*(1), 94.
41. Pearce, D. W., & Turner, R. K., (1989). *Economics of Natural Resources and the Environment* (Vol. 1, pp. 1–23). Johns Hopkins University Press.
42. Peigné, J., & Girardin, P., (2004). Environmental impacts of farm-scale composting practices. *Water, Air, and Soil Pollution, 153*, 45–68.

https://doi.org/10.1023/B:WATE.0000019932.04020.b6 (accessed on 30 May, 2020).

43. Province of Ontario, (2017). *Strategy for a Waste-free Ontario—Building the Circular Economy* (Vol.1, pp. 1–22). Toronto, ON: Queen's Printer for Ontario. https://files.ontario.ca/finalstrategywastefreeont_eng_aoda1_final-s.pdf (accessed on 30 May, 2020).
44. Raviv, M., (2010). The use of mycorrhiza in organically-grown crops under semi-arid conditions: A review of benefits, constraints and future challenges. *Symbiosis, 52*(2–3), 65–74.
45. Rosenfeld, P. E., & Suffet, I. H., (2004). Understanding odorants associated with compost, biomass facilities, and the land application of biosolids. *Water Science and Technology, 49*(9), 193–199.
46. Rosenfeld, P. E., Clark, J. J. J., Hensley, A. R., & Suffet, I. H., (2007). The use of an odour wheel classification for the evaluation of human health risk criteria for compost facilities. *Water Science and Technology, 55*(5), 345–357.
47. Sánchez, O. J., Ospina, D. A., & Montoya, S., (2017). Compost supplementation with nutrients and microorganisms in composting process. *Waste Management, 69*, 136–153. https://doi.org/10.1016/j.wasman.2017.08.012 (accessed on 30 May, 2020).
48. Sánchez-Monedero, M. A., Serramiá, N., Civantos, C. G. O., Fernández-Hernández, A., & Roig, A., (2010). Greenhouse gas emissions during composting of two-phase olive mill wastes with different agroindustrial by-products. *Chemosphere, 81*(1), 18–25.
49. Scheuerell, S. J., & Mahaffee, W. F., (2006). Variability associated with suppression of gray mold (*Botrytis cinerea*) on geranium by foliar applications of nonaerated and aerated compost teas. *Plant Disease, 90*, 1201–1208. http://doi.org/10.1016/j.wasman.2017.08.012 (accessed on 30 May, 2020).
50. Smil, V., (2001). *Enriching the Earth: Fritz Haber, Carl Bosch, and the Transformation of World Food* (Vol. 1, p. 33). Boston: MIT Press.
51. Stahel, W. R., (2016). Circular economy: A new relationship with our goods and materials would save resources and energy and create local jobs. *Nature, 531(*7595), 435–438.
52. Stewart, B. A., Robinson, C. A., & Parker, D. B., (2000). Examples and case studies of beneficial reuse of beef cattle by-products. *Land Application of Agricultural, Industrial, and Municipal By-Products, 6*,

387–407.

53. USDA-NRCS, (2012). *Farming in the 21st Century: A Practical Approach to Improve Soil Health* (Vol. 1, pp. 1–30). Washington, D.C: National Soil Health and Sustainability Team, Natural Resources Conservation Service, United States Department of Agriculture. https://www.nrcs.usda.gov/wps/portal/nrcs/main/soils/health (accessed on 30 May, 2020).
54. Van, E. H., (2016). *What Is Soil Health?* (pp. 1–30). Fact Sheet Number 16-02. Soil Health Manual Series. Soil Health Laboratory, Cornell University, Ithaca, NY.
55. Walker, P., Williams, D., & Waliczek, T. M., (2006). An analysis of the horticulture industry as a potential value-added market for compost. *Compost Science and Utilization*, *14*(1), 23–31.
56. Walther, E., Ferrier, R., Bennacer, R., et al. (2017). Heat recovery in compost piles for building applications. *Thermal Science, 21*, 775–784. https://doi.org/10.2298/TSCI160411299W (accessed on 30 May, 2020).
57. Wei, Y., Li, J., Shi, D., et al., (2017). Environmental challenges impeding the composting of biodegradable municipal solid waste: A critical review. *Resources, Conservation and Recycling, 122*, 51–65. https://doi.org/10.1016/j.resconrec.2017.01.024 (accessed on 30 May, 2020).
58. Wolsink, M., (1994). Entanglement of interests and motives: Assumptions behind the NIMBY-theory on facility siting. *Urban Studies*, *31*(6), 851–866.
59. Wong, J. W., & Fang, M., (2000). Effects of lime addition on sewage sludge composting process. *Water Research*, *34*(15), 3691–3698.
60. Wu, X., Wei, Y., Zheng, J., Zhao, X., & Zhong, W., (2011). The behavior of tetracyclines and their degradation products during swine manure composting. *Bioresource Technology*, *102*(10), 5924–5931.
61. Zinati, G. M., (2005). Compost in the 20th century: A tool to control plant diseases in nursery and vegetable crops. *Hort. Technology*, *15*(1), 61–66.

# 7

# *Introduction to Fishery by-Products*

## CONTENTS

## 7.1. INTRODUCTION

As by-catch, there is a huge quantity of very small fish landed which as fresh fish do not find a ready market. The filleting and processing industries are known to be the large quantities of fisheries waste. All these turn out to be good sources of good quality mineral, fat, protein, etc. In our country, the current fish-processing is decades old (Kim and Lee, 2015). More or less in the fish processing industry, the impressive increase in the export of marine development and products has been based on prawns and fish, which compose about more than 50% of the total marine catch. In several other fish by-products and products however, this current trend is not so obvious. Fish meal is one of the significant by-products of commercial value, from underutilized by-catch species and fish species it mainly produced (Green et al., 1977; Marali et al., 2013) (Figure 7.1).

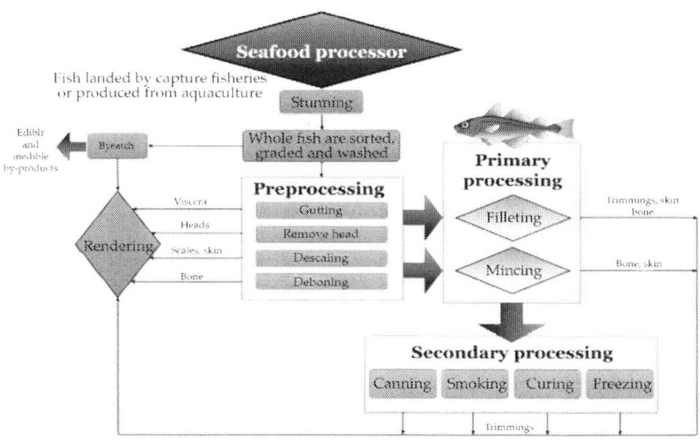


**Figure 7.1:** Processing of fishery for production of different products.

*Source:https://www.sciencedirect.com/science/article/pii/B9780128023914000033.*

## 7.2. MARINE HABITAT

The salinity contained by marine waters ranges from 33 to 37 PSU, and in-depth, it can range up to 11,000 m in the Mariana Trench, which is the deepest ocean bench. Aphotic and photic are two zones in which oceans are divided. Penetration of light up to 200 m depth depends on the volume of turbidity, present in water. Lower penetration of light <1 m is received by few areas near the coasts; this happens because in the water, suspended particulate matter has a high level. From a microbiological point of view, based on habitats, the marine environment is classified by four important habitats. The air-water interface at the surface is cited as the neuston, the planktonic habitat or water column is called the pelagic zone and on the basis of water column depth, this zone is further subdivided. In the upper 100 m of the water column, the habitat that is found is named as the epipelagic zone, for instance, the photic zone. Mostly photosynthetic organisms are present in this zone. A sea-sediment interface is referred, by the benthopelagic zone (benthos). The third one is epibiotic habitat, referring to attached communities occurring surfaces. Endobiotic habitat is the fourth one, it refers to those organisms which exist in the tissues of other organisms (of larger ones), for instance, i.e., fish. Vibrio Fischer is an example of such an endobiotically living bacterium, it uses quorum and luminescence sensing in its distinctive lifestyle. Diverse microbial habitats are contained by the oceans; the diversity is extremely low below the neuston region and is highest in it. Large numbers of microbes are contained by seawater, though it is far off from impacts caused by humans, however, in the coastal areas, a greater concentration of microbes is found. Many plankton and bacteria occur in deep ocean water, as well as in surface water. For sheathing their genetic material and to prepare copies, bacteria or other cells are required by the viruses. Scientific studies suggest that a small amount of seawater (a teaspoon even) contains 10–100 million viruses.

## 7.3. MARINE ECOSYSTEMS

A realm for multiple different species is formed by marine ecosystems, these species range to large marine mammals from planktonic organisms that sets up the basis of the marine food web. Marine ecosystems are explored by different species, with the aim of getting food and seeking shelter from predators. For the comprehensive health of both terrestrial and marine environments, they are extremely important. Coastal habitats are the ones having a look from above the water level or above the limit of the spring high tide in nontidal waters. Coastal habitat's closeness to the sea involves several habitats such as beaches, supralittoral habitats, coastal dunes, cliffs, and sandy shores. Coastal habitats solely account for around 30% of total marine biological productivity. For humans, productivity, and assortment are also important. A considerable source of income and food is dispensed by these habitats. Species serving as fertilizers, animal feed, additives in cosmetics and food are also supported by them. Seagrasses and mangroves are habitats that protect the coastlines from erosion and wave action, while other areas act as filtering systems or provide sediment sinks. Significant damage is kindled by increased human activities and by these activities, marine biodiversity is also under serious threat. These activities might be pollution, coastal development, overfishing, or the introduction of exotic species. Hence, for ensuring the prevention and safety of these ecosystems, serious cautions for conservation and safeguarding of marine ecosystems must be exercised.

## 7.4. HARVESTING AND HANDLING OF FINFISH

The procedure employed for harvesting affects the microbial load and the primary quality of fresh finfish. Hence, quality maintenance must be initiated at this stage. Improper handling during harvesting will be harmful to the ensuing shelf life and quality at the retail level. Commercial harvesting of finfish involves a wide range of procedures and fishing gear including hook and line techniques, numerous kinds of nets, and traps and barriers. This increase is a result of the drag along the ocean's base, which makes the mud move up and confines the fish, leading the gut contents to be indicated. Commonly, during trawling lower tows produce lower quality. During heavy fishing the end of the cod becomes extremely full causing the catch, which may have been lifeless for hours, to become damaged and crumbled due to compression. Additionally, larger catches require appropriate longer stowing. The fish are then, exposed to the abuse of sliding on the deck of the boat as well as the environment's temperature and sunlight. Hence, quality is negatively affected because of fishing for quantity.

The postharvest quality also seems to be affected by the stress level experienced by the fish after they are caught, just before dying. Fish, like mackerel and tuna, are generally very active and may become excited and die in a hyperactive state when harvested through purse-seining. Other species differ from tuna based on the natural physiological attributes possessed by it. With a burst speed for yellow fin stated to be as 21 body lengths per second by *(T. albacores),* they are one of the fast-moving fish. Tuna exhibit a very high metabolic rate with some species being capable enough to modify their body temperature. A serious defect of the flesh, a condition known as burnt flesh is experienced when tuna is caught in an extremely stressed state, causing the lactic acid present in the muscles to combine with the increased muscular temperature. This results in the production of an acidic flavor, accompanied by a metallic after taste in the flesh which no longer seems to be bright red. The tuna is still adequate to be canned; however, it is unsatisfactory for the extremely extensive Japanese sashimi market. Therefore, to reduce stress and to regulate tuna's postharvest quality, the long lining technique is employed for desirable harvesting. The impact of stress has also been observed in other species of fish. Salmon die after a tiring struggle when harvested through gill netting. Consequently, the process of rigor mortis is triggered as well as initial signs of deterioration that take place during icing. By harvesting through the method of hook and line, the fish are rapidly moved aboard the vessel and killed, decreasing the stress and the quality deterioration related to it. This idea of "clean kill" quite famous in the slaughter of poultry and livestock. Research on the harvesting of Atlantic cod suggested that the degree of sniggling largely affected the flesh's caloric and protein content by decreasing the moisture. A concurrent analysis by the same researchers implied that the technique used for capturing Atlantic cod was more important than the time of the season in its influence on sensory quality.

## 7.5. FISHING VESSEL SANITATION

The cleanliness of food handling equipment and food-contact surfaces is commonly given great importance in the food industry. However, onboard fishing vessel practices by the food industry are not appropriate. Counts aboard the vessels are very high due to the natural presence of high bacterial load on fish. Shepherd and Jackson (2013) stated the logos psychrotrophic counts/inch of the contact surfaces as hand gaffs, 8.2 organisms; decks, 7.3; fish hold, 6.5; and shovels/pushes, 7.7. Analyzers described that even though the sanitary environment amidst the commercial fishing vessel could be made better; it is not harmful to a well-iced catch. This recommendation has coincided with the studies of (Tacon and Metian, 2008). Proper maintenance of sanitation is required to prevent cross-contamination of the fish stored in dirty ice from the past trips or with bilge oil even if it may not benefit the shelf life and quality of whole fish which are well-iced. The execution of onboard processing can promote reduced shelf life and quality by exposing the earlier sterile flesh to an immense amount of spoilage bacteria due to an unsanitary environment. Fish holds should be built or modified to make them facilitate sanitary practices. Wood, for instance, cannot be sanitized adequately.

## 7.6. FISHMEAL

Fishmeal is usually termed as a commonly used livestock feed supplement. Fishmeal contains three of the vital amino acids in which the animal bodies are unable to synthesize, cysteine, and methionine, the high-quality protein that has high levels of lysine, and also an unrivaled constituent of feedstuff be made through this. Fishmeal is also a source of Vitamin-B like pantothenic acid, chlorine, riboflavin, cyanocobalamin (B12) and niacin. It is a source of some trace elements and also rich in minerals like copper, phosphorus iron, and calcium. From canning and filleting operations, by cooking, pressing, drying, and grinding skeletal remains it is produced accompanied by adhering proteinaceous tissues of fish. Or via processing whole miscellaneous fish which along with prawns mainly caught, which contain Ribbonfish, sole, silver-bellies, Jewfish, and the alike. Because of variations in raw materials used the composition of fish meal differs significantly, conditions employed and processing methods. For fish meal the major raw material is abundant but, on the west-coast sporadic catches of oil-sardine (Tacon and Metian, 2008; Shepherd and Jackson, 2013) (Figure 7.2).

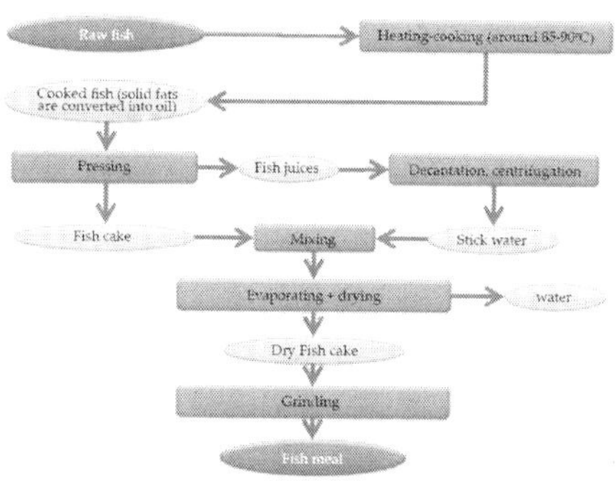


**Figure 7.2:** Flowchart for production of fishmeal.

*Source: https://www.feedipedia.org/node/208.*

### 7.6.1. Raw Materials

To manufacture fishmeal high-fat such as menhaden, sardines, herring anchovies, etc. are commonly used as raw materials. From shrimp small by catch fish searching usually not marketable as fresh fish because of several reasons like bony nature, very small size, etc. also can be used.

Juveniles of waste and commercially important fish from filleting plants and fish processing, carcasses of fish like shark, and cannery wastes other fish wastes for fish oil and meal production are also used as raw material. In India, common fish meal production was from sun-dried fish gathered from several drying centers all along the coast, and as manure, the product was mainly used. From the very beginning of this industry better quality fish meal has been an important item of export. The importance of increasing quality for good use was felt, and the Ministry of Agriculture and Food has, in 1959, regarding quality fish meal laid down specifications. Later for the fish meal as livestock feed for enabling proper quality control the Bureau of Indian Standards (BIS) has brought out the specification (IS: 43071967) (Torstensen et al., 2008).

### 7.6.2. Composition of Fish Meal

Usually the range of proximate analysis gained by following ways:

Ash, 12–33%; Fat: 05–10%; Protein: 50–70%; moisture, 06–10%.

Fish meal is rich in all minerals mainly calcium and phosphorus, Vitamins B and the important amino acids.

By the two general processes fish can be reduced to fish meal, i.e., wet-rendering, and dry-rendering.

### 7.6.3. Wet-Rendering

The process which is usually applied to offal or fatty fish where simultaneous production of fish-body oil and fish meal is envisioned is known as wet reduction and wet rendering. The process includes Grinding-cooking to make softer bones and flesh and to release the oil, pressing to eject oil and liquor, puffing press-cake, drying, grinding, and packing meal with 8% moisture, to eradicate suspended particles and to separate concentrating stick and oil-water centrifuging press liquor is used.

The process needs expanded equipment and is usually a continuous one and so adjustable to the reduction of large quantities of fish.

### 7.6.4. Dry-Rendering

The process which is employed to process fishmeal from non-oily fish, for example, ribbonfish, Jewfish, silver-bellies, carcasses of shark, anchovies, and sole, filleting waste and fish offal is known as dry reduction or dry rendering. In this process, it is dried to the moisture content of 10% and pounded if the quantity of fish processed is very low. A steam-jacketed cooker-dryer furnished with the power-driven stirring device is used if quantity to be handled is appropriately large. Being a batch operation, the process will have just limited labor costs and capacity will be high. However, in meal water-soluble materials are remained.

## 7.6.5. Present Status

A sign of revival is being shown by the organized fish meal industry during recent years. In our country with an installed capacity of 175 tonnes/day, there were 7 such plants in 1972. Now around 12 modern fish meal plants of wet-reduction type at several fishing centers are in operation.

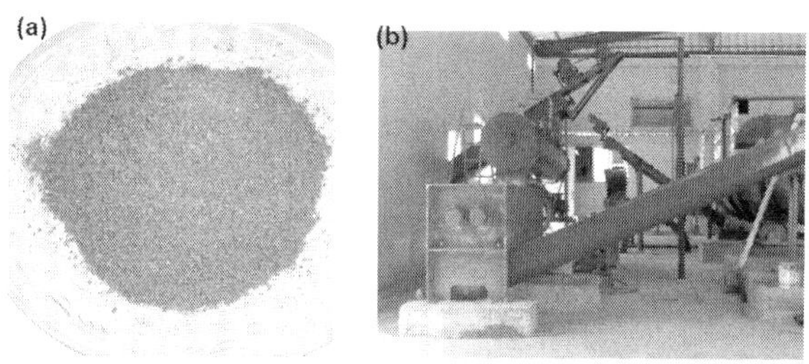


**Figure 7.3:** (a) Fishmeal and (b) fishmeal preparation plant.

*Source:https://www.researchgate.net/publication/259356854_Fishery_By-Products*

## 7.6.6. Uses

In the livestock feed, the main application of fish meal is as an ingredient. At the following levels in the ratios of poultry and animals fish meal is frequently supplemented:

- Sheep: 45–91 g/day 45.4 kg live weight;
- Pig: 113–127 g/day according to weight;
- Cattle: 907/ g/day/kg live weight; and
- Poultry: Not more than 5% for chicks and more than 10% of the total ration for hens.

Due to the growing aquaculture activities, another vital use of fish meal is in fish and shrimp feeds that have become a very significant application. From 10–70% the amount of fish meal used in these feeds differs. Fish meal stabilized with antioxidant or low in fat is ideal in aquaculture feeds.

## 7.7. FISH OILS

There are two types of fish oil, body oil, and liver oil. For therapeutic purposes in the treatment of vitamin D and A deficiencies, fish liver oils are generally used. Due to the content of polyunsaturated fatty acids (PUFA) body oils have recently gained attention, mainly in the control of heart ailments in humans η3 PUFA are used (Holub and Holub, 2004; Virtanen et al., 2005) (Figure 7.5).In the 18th century, the therapeutic values of fish-liver oil were explored, and fish liver oil became a general therapeutic product. In specific fish-liver oils, both vitamin D and A are found. From shark, haddock, and cod the most significant fish-liver oils are obtained. Tuna and halibut livers are also rich bases of vitamin D and A. The presence of vitamins, fat content and weight of liver reliant on many factors such as sex, age, species, area from where it was caught, stage of spawning and nutritional status (Mori et al., 1994) (Figure 7.4)

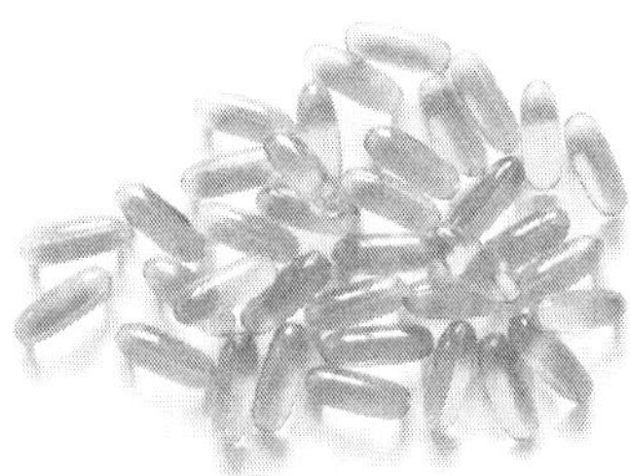

**Figure 7.4:** Fish liver oil capsules.

**Figure 7.5:** Fish body oil capsules.

*Source: https://www.researchgate.net/publication/259356854_Fishery_By-Products.*

## 7.8. FISH SILAGE

When fish is available for numerous reasons which cannot be used for direct consumption, for the production of the fish meal it is used, which as an animal-feed has got a ready market. If the material is present in such a place where there is no reasonable transport to the nearest plant and no fish meal plant, due to fish odors there may be severe limitations on fish meal production, for alternate processes one has to look. It is useful to go for silage production when animal farms are extremely near to landing stage centers (Dapkevicius et al., 2000; Gildberg, 2004). The product which is made from parts of fishes or whole fish no other material has been added to which, except an acid, and through enzymes, already present in the fish liquefaction of fish is brought is known as fish silage. By treating it with formic acid (organic acid) or sulfuric acid (mineral acid) fish silage can be made from parts of fish or whole fish. Ensiling can also be accomplished by lactic acid that is formed by fermentation of sugar using a starter culture of lactic acid bacteria. Molasses is added to the fish offal or fish as the source of sugar. With a malty odor, the product is a stable liquid, which comprises all water present in the original material and has brilliant storage characteristics. It is a very easy process and needs little capital apparatus mainly if non-oily fishes are used. The oil separation is required for the use of oily fish, for which temperature is raised to 70–900°C and by centrifugation or decantation, floating oil is separated (Stone and Hardy, 1986) (Figure 7.6).

**Figure 7.6:** Fish silage.

*Source: https://www.researchgate.net/publication/259356854_Fishery_By-Products.*

As cattle feed fish silage is used. Either the decanted liquid portion of the whole mass can be used. the Silage is mixed with rice bulk or other feed ingredients when solid feed is required.

## 7.9. FISH PROTEIN CONCENTRATE (FPC)

Any even fish preparation, planned for human consumption, wherein protein is more concentrated than in original fish is known as Fish protein concentrate (FPC) (Cheftel et al., 1971).

Following three types of FAO is defined by FPC:

1. Type A: It is an almost tasteless and odorless powder that has a fat content of 0.75%;
2. Type B: Powder that has no exact limits as to odor of flavor, but certainly having a maximum fat content of 3% and fishy flavor; and
3. Type C: Under reasonably hygienic conditions normal fish meal produced.

**Figure 7.7:** Fish Protein concentrate.

*Source: https://www.researchgate.net/publication/259356854_Fishery_By-Products.*

FPC is as such is not appreciated for consumption though it is intended for human consumption. As a protein enhancement in the human diet, it is therefore combined. In biscuit and bread, the FPC level of 5–10% is considered a satisfactory limit (Chen et al., 1975; Hevia et al., 1976) (Figure 7.7).

## 7.10. CHITIN

After the cellulose chitin is the most abundant organic compound on earth. It is an inelastic nitrogenous polysaccharide, hard, white extensively circulated in the exoskeleton of lobster, insects, shrimp, and crab and also in the invertebrate's internal structures, for example, squid. Head waste and shrimp shells in India constitute the single main source of chitin (Kumari et al., 2015).

### 7.10.1. Processing

There are two steps are involved in the production of chitin from crab waste or shrimp viz., by using a mineral acid demineralization of the waste after that by using a dilute solution of caustic soda deproteinization of the demineralized mass. If dry shell waste is used the bulk of the mass is reduces by the initial demineralization, which makes the deproteinization step simpler. With dilute hydrochloric acid (1.2 N) the shell waste is stirred well till it becomes soft. Demineralization may take up to one hour for the conclusion and is shown by the slowdown of effervescence. The liquid is poured and the mass is washed through the water till free of acid. In a 5% solution of caustic soda in water for a few minutes, the demineralized mass is boiled. The liquid portion comprises of the dissolved protein is clean off and the rest is washed with water till free of alkali. Under a hot air drier or in sun the wet mass is dried. Chitin is produced. Before bagging it is usually pulverized to the required mesh size.

### 7.10.2. Uses of Chitin

For various purposes, chitin is used. In birds and animals' chitin is a growth promoter and in their feed is used as ingredients. The acetylated chitin, chitosan, is a usual flavoring material which can form films and ionic bonds (Madhavan and Ramachandran Nair, 1974; Defoirdt et al., 2010).

## 7.11. BIVALVE MOLLUSCAN SHELLFISH

Most of the sea food-associated illnesses in the US are related to the ingesting of bivalve molluscan shellfish. This disproportionate occurrence of the disease is linked to the biology of animals, the quality of water sin which the diseasesgrow, the handling methods after harvest, and the point that they are of ten consumed raw. This chapter will reflectall these influences from amicrobiological viewpoint.

The majorityof knowledgeon the micro biology of bivalve molluscan shellfish has been established comparative to public health traits of the shellfish industry. This knowledge has been inferred into state and federal regulations that managethatindustry. To developanunderstandingofthe shellfishindustryandthe needforregulations, refer to themostcurrent editionof the NationalShellfish Sanitation Program, Manual of Operations, Parts I and II.

## 7.11.1. Bivalve Processing

Raw processing of the bivalves is restricted to the elimination of the meat from the shell, washing of the meat with potable water, and packing the meatin containers. Sealed containers are kept in boxes and then covered with crumpledice for freezing and transport.

Shucking involves compelling a knife amongst the 2 valves of mollusc shell and slicing the adductor muscle where it is fixed to one valve. Careis normally taken not to slicein to the meat. The knife is utilized to slice the adductor muscle, therefore, freeing themeat.

Washing of the meati sachieved by keeping the monthe stand-supported punctured tray known as the skimmer and spraying with the potable water. An alternate technique of washing is blowing. Meats are kept in the tank of water and disturbed by the introductiono fair. Bothwashingmethods eradicatemucus, mud, shellfragments, and pseudofecesfromthemeats. Meats are thoroughly drained before packing.

Shucking and washing of the oyster meat under sanitary conditions decreases the SPCandthe25°C APC (aerobicplatecount) ofthemeatbyaround1log. Conversely, thenumber ofindicatorbacteria, A. hydrophilia, and vibriosmightnot beconsiderablychangedbyprocessing

Shellfish processors at times add chlorine (Cl) to the water in blowing tank in an effort to decrease the bacterial load on the bivalve meats. Kandra et al. (2012) validated that blowing oysters in 150 parts per million chlorinewaterfor20 to 30 minutescausedthereductioninaerobicplatecount andchangedthe speciescompositionofoystermicroflora. V*ibrio*sp. was actuallyleast disturbed the chlorine (Cl) treatment and amplified from 23.0 to 63.7% of the population.

Preservation of good hygiene in the processing plant averts the bivalve meat from becoming polluted during raw processing. Quick freezing of the shucked meat to near 0°C retards spoilage.

## 7.11.2. Heat Processing

### *7.11.2.1. Heat Shock*

Hand shucking of the raw bivalves can normally be made simpler by subjecting bivalves to slight heat treatment just before shucking. The temperature and time of the heat treatment are dependent on the bivalve size and species. This process doesn't eradicate the bivalvebut causes bivalve to easeitsadductor muscle hence the shell scan be easily opened. Further more, heat shock causes are duction in the number of bacteria in the shucked meat.Heat shock is utilized to facilitate meat elimination from bivalves in order to be canned. Inthissituation, thetreatment of heat is more severe and migh tcomprise steaming under pressure for many minutes. The juice lost by bivalves during the heat shock is gathered and utilized incanning.

### *7.11.2.2. Canning*

Canning is utilized as a long-term preservation approach. Procedures for canning several bivalve species have been printed and if suggested processing temperatures andtimesarefollowed, some of the microbiological issues areencountered.

### *7.11.2.3. Pasteurization*

Pasteurization has been utilized on the experimental basis to extend the shelf life of the raw oysters without reducing their flavor and texture. Kandra et al. (2012) discovered that pasteurization at 72°C for eight minutes in the reportable pouch occasioned in the product of higher microbiological quality that didn't considerably vary from fresh oysters in appearance, flavor, odor, or texture. The facultativeanaerobic and aerobicplatecountsstayedlow for three months when the past eurized oysters were kept at 0.5°C. However, resultsweren'treliable among the batches of pasteurized oysters, see mingly be cause of the difference in the species and amounts of bacteria in an individual oyster before processing.Instantaneously following pasteurization, most of the bacteria in oysters were *Bacillus* sp. During storage for around five months, *Bacillus* sp. sustained to govern the aerobic microflora while *Clostridium*, C*orynebacterium*, *Peptostreptococcus*, *Listeria*, and *Staphylococcus* appeared in the facultativeanaerobicmicroflora. Although the microflora of oysters was considerably improved by the heat treatment, the usual phases of spoilage were observed, comprising an increase in gas formation, acidity, and proteolysis.

### *7.11.2.4. Freezing*

Freezing has been utilized efficiently to hinder the microbiological spoilage of various fish and shellfish. If the products are glazed properly to avoid dehydration and placed under the constant temperature of 6.6°C, they will possess a shelf life of several months.

Freezing isn't meant to kill bacteria in the fishery-products, but instead to avoid them from growing, metabolizing, and causing spoilage. However, the freezing procedure normally brings about an early reduction in the number of bacteria followed by a steady decline during storage. The persistence of bacteria during freezing and following storage is dependent on the bacteria species.

Freezing considerably lowered the SPC of the Eastern oysters, but after melting their shelf life was similar to oysters that hadn't been frozen as measured by SPC.

## 7.12. PRAWN HEAD MEAL AND SHELL WASTE

The shell and head of prawn/shrimp and another crustacean where prawn is rich form the major fishery waste. In countries like Thailand, India, Mexico, South Africa, Philippines, Malaysia, and the USA, lobster, crab, and shrimp turn out massive quantities of shell waste; the discarding of which creates a serious problem. This waste has a good percentage of chitin and protein other than minerals. The protein can be removed together with flavor-bearing compounds and converted into shrimp excerpts having the potential to be used as a natural flavoring material (Kandra et al., 2012).

To minimize spoilage shell waste and prawn head intended for extraction of protein and conversion to shrimp extract should be collected fresh and stored in ice. To eradicate all adhering dirt, and other unnecessary matters it should be washed in clean water. By boiling with 0.5% sodium hydroxide solution in water the protein is then extracted. To separate remaining shell the boiled mass is filtered through an appropriate sieve. With HCl (hydrochloric acid) the filtrate is neutralized initially, and with acetic acid to the end to a pH 6.8 to 7.0. By boiling in open steam jacketed kettle with the moisture content of about 35 to 45% to a semi-solid mass the neutralized filtrate is concentrated. The paste formed is usually known as shrimp extract. It contains on average 40% partially hydrolyzed protein, 40% moisture, and 10% minerals. The yield is 20% of the shells and fresh heads (Adeyeye and Adubiaro, 2004; Amar et al., 2006).

## 7.13. VALORIZATION OF FISHERY BY-PRODUCTS

In the last few years, the fishery wastes problem has increased and become a universal concern that is being affected by numerous operational factors, technical, and biological along with socioeconomic drivers (Kim and Mendis, 2006; Arvanitoyannis and Kassaveti, 2008). Many by-catch products having low or no commercial value or fish species, damaged or undersized commercial species along with species of commercial value but not caught in enough amounts to warrant sale are termed as the "fish wastes." The use of fish, however as feed cannot be controlled only through fishery market forces and, in contrast, the need for aquaculture development and responsible fisheries has just been emphasized to preserve aquatic biodiversity (FAO, 2011).

It has been projected that over 50% of fish tissues are considered as the "wastes" like viscera, heads, skin, and fins and they are discarded. Discards from the world's fisheries every year cross 20 million tons which is about 25% of the entire production of marine fishery catch and contains fish processing wastes, by-products, and "non-target" species. A total of ca. 5.2 million tons represented by the discards in the European Union per year (Rustad, 2003; Mahro and Timm, 2007) (Figure 7.8).

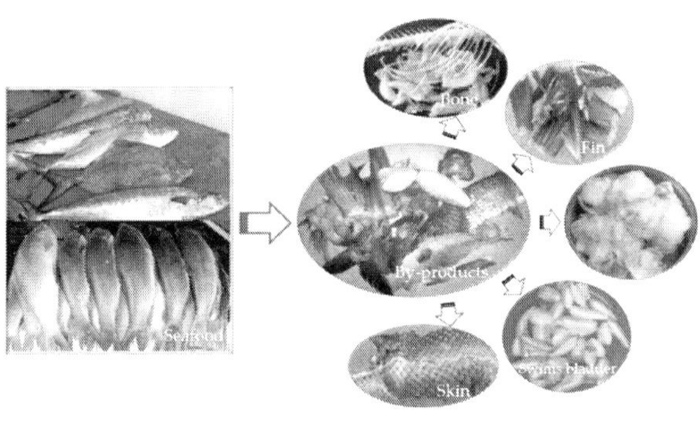


**Figure 7.8:** Classification of different fish-By-products for valorization.

*Source: https://www.sciencedirect.com/science/article/pii/S1 466856 4163004 80.*

For the fishing areas the composition and the amount in types of fishery wastes are different, the total amount of discarding, therefore, is extremely variable (Rochet and Trenkel, 2005; Davies et al., 2009). Gonçalves et al. (2007) reported that in the Mediterranean Sea, significant amounts of trammel net (a fishing net that has large mesh at the edges and smaller mesh in the middle) discards, with a total discard rate ranging from 15 to 49% for Greece to Portugal of the total of 137 discarded species (79.7% of the total). Furthermore, it is estimated that the discards of fish processing industry would reach up to 75% of the total volume of products (Olsen et al., 2014).

The environmental impact is another important feature that fish wastes might have on aquatic ecosystems, the biodiversity of the benthic and community assembles as the release of organic wastes change significantly (Vezzulli et al., 2002, 2008). To reduce this pollution source and to look for the best ways to solve this problem the management of fish discards involves several aspects. The new Common Fisheries Policy (CFP) for example, in Europe, aims to decrease the discard rate.

Along with the enormous economic loss, fish waste for farmed fish could be planned as a source of feed. Last year the increasing trend shown by aquaculture production has been documented by various studies (FAO, 2012), however, by a significant blockage, the future aquaculture development is being constrained. On the availability of fish oil and meal, aquaculture productions are reliant on, which are for fish feeds to become an important lipid and protein source (Jackson, 2009). the total costs of production could significantly be affected by the production of aqua feeds; on average, there is 4 to 6 kg of fish required for the production of 1 kg of fish meal, whereas the 10–50 kg of fish is required for the production of 1 kg of fish oil (Péron et al., 2010; Shepherd and Jackson, 2013). It has been estimated that the aquaculture field consumed about 80% of fish oil and 60% fish meal of the world total production yearly and this percentage is likely to increase to warrant future aquaculture growth (Naylor et al., 2009; Tacon et al., 2011). Fish oil and fish meal signify the most valued products got from marine by-products and are not useful for human consumption (Ferraro et al., 2010). For the production of fish oil and fish meal in 2012, 15 million tons of fish were caught. Per year globally among 5.5 to 7.5 million tons of fish oil are produced but in the future, further growth of the intensive aquaculture productions will be limited by the availability of fish oil and fishmeal (FAO, 2002, 2009; NOAA/USDA Alternative Feeds Initiative, 2011).

Chile and Peru are the most important countries for the production of fish oil which contributes to 13% and 52%, respectively, of the total world production. Norway and Iceland can fulfill the whole European need and contribute to 7% of the world's fish oil production (EU, 2003; European Parliament, 2004). The fishery by-products could be used as the alternative source for the supply of fish oil and fish meal and the fishing pressure on the species selected for oil and meal production could be reduced and, simultaneously, for sustainable aquaculture productions it could contribute (Asche and Tveterås, 2004; EU, 2013).

Despite the low traditional value that assigned to fishery by-products, a great number of bioactive compounds with wide biotechnological and pharmaceutical applications could be produced from this huge mass of under-utilized/unused resources, like proteins (collagen, enzymes), protein hydrolysates, chitin, astaxanthin, lipids (Ghaly et al., 2013; Khora, 2013). Among the bioactive compounds take out from fishery by-products and wastes, oils, and proteins rich in PUFA, particularly docosahexaenoic acid (DHA) and eicosapentaenoic acid (EPA) are mainly interesting for their possible use as fish feed components along with their high marketable value, (Hardy et al., 2005). As fishery wastes are rich in high-quality nutrients, they have a great potential to utilize and convert a large portion of these useful products in marine bioprocess industry.In terms of environmental and management impacts, on the other hand, fishery by-products and discards may guide to significant problems. Particular importance in many countries is given to discover the possibility of using by-products of fishing, traditional fishing, and aquaculture instead of facing the problem of their disposal. For the valorization of fishery discards and by-products at a regional or local level, the use for aqua feeds production of the wastes from the fish processing and fishing industry could be an important tool (Kotzamanis et al., 2001; Falch et al., 2006).From European fleets, fishery discards are not insignificant, and their decrease is a socioeconomic, ecological, and moral imperative. By significant fishing fleets, the countries bordering on the Mediterranean Sea are characterized; as a source of aquafeeds along with bioactive compounds based on these considerations, this basin may play a significant role. To remove discards and decrease unwanted by-catches in European fisheries the EU (European Union) has just launched a joint policy, but an execution of the EU (European Union) Regulation for the decrease of fish wastes is still needed (Kim and Wijesekaraa, 2010). By establishing technologies to transform and increase fish wastes in an economic resource the goal of decreasing fishery discards can be accomplished, for instance by developing techniques of concentration and extraction of the bioactive compounds they defining and containing policies for their use for fish oil and meal for animal feeding. Furthermore, an ecosystem method instead without an important interference on the selectivity of the used fishing equipment specifies that policy for the effective decrease is not possible (Hordur et al., 2000; Harnedy and FitzGerald, 2012).

## REFERENCES

1. Adeyeye, E. I., & Adubiaro, H. O., (2004). Chemical composition of shell and flesh of three prawn samples from Lagos lagoon. *Journal of the Science of Food and Agriculture*, *84*(5), 411–414.
2. Amar, B., Philip, R., & Bright Singh, I. S., (2006). Efficacy of fermented prawn shell waste as a feed ingredient for Indian white prawn, *Fenneropenaeus indicus. Aquaculture Nutrition*, *12*(6), 433–442.
3. Arvanitoyannis, I. S., & Kassaveti, A., (2008). Fish industry waste: Treatments, environmental impacts, current and potential uses. *International J. Food Sci. Technol, 43*, 726–745.
4. Asche, F., & Tveterås, S., (2004). On the relationship between aquaculture and reduction fisheries. *J Agricult Econom., 55*(2), 245–265.
5. Cheftel, C., Ahern, M., Wang, D. I., & Tannenbaum, S. R., (1971). Enzymic solubilization of fish protein concentrate: Batch studies applicable to continuous enzyme recycling processes. *Journal of Agricultural and Food Chemistry*, *19*(1), 155–161.
6. Chen, L. F., Richardson, T., & Amundson, C. H., (1975). Some functional properties of succinylated proteins from fish protein concentrate. *Journal of Milk and Food Technology*, *38*(2), 89–93.
7. Dapkevicius, M. L. E., Nout, M. R., Rombouts, F. M., Houben, J. H., & Wymenga, W., (2000). Biogenic amine formation and degradation by potential fish silage starter microorganisms. *International Journal of Food Microbiology*, *57*(1–2), 107–114.
8. Davies, R. W. D., Cripps, S. J., Nickson, A., & Porter, G., (2009). Defining and estimating global marine fisheries by catch. *Mar Policy, 33*(4), 661–672.
9. Defoirdt, T., Darshanee, R. H. A., Karunasagar, I., Boon, N., & Bossier, P., (2010). Quorum sensing negatively regulates chitinase in *Vibrio harveyi. Environmental Microbiology Reports*, *2*(1), 44–49.
10. EU, (2003). The use of fish by-products in aquaculture. Health and consumer protection. Directorate General. Ed. *Report of the Scientific Committee on Animal Health and Animal Welfare, 1*, 88–93.
11. EU, (2013). *Strategic Guidelines for the Sustainable Development of EU Aquaculture* (Vol. 1, pp. 1–14). Brussels.
12. European Parliament, (2004). *The Fish Meal and Fish Oil Industry:*

*Its Role in the Common Fisheries Policy* (1, 140–148). Directorate General for research, Luxembourg.

13. Falch, E., Rustad, T., & Aursand, M., (2006). By-products from gadiform species as a raw material for production of marine lipids as an ingredient for food and feed. *Process Biochem, 41*, 666674.
14. FAO, (2002). *Use of Fishmeal and Fish Oil in Aqua Feeds: Further Thoughts on the Fishmeal Trap* (pp 55–61). N.B.New & U.N. Wijkström. FAO Fisheries Circular No. 975.
15. FAO, (2009).In: Hasan M. R., & Halwart, M., (eds.), *Fish as Feed Inputs for Aquaculture—Practices, Sustainability and Implications* (pp. 420–426). FAO Fisheries Aquacult Tech Paper, 518.
16. FAO, (2011). Aquaculture development. 5. Use of wild fish as feed in aquaculture. *FAO Technical Guidelines for Responsible Fisheries, 5*(Suppl. 5), 70–79.
17. FAO, (2012). *The State of the World Fisheries and Aquaculture 2012* (p. 209). FAO, Rome.
18. Ferraro, V., Cruz, I. B., Ferreira, J. R., Malcata, F. X., Pintado, M. E., & Castro, P. M. L., (2010). Valorisation of natural extracts from marine source focused on marine by-products: A review. *Food Res. Int, 43*, 2221–2233.
19. Ghaly, A. E., Ramakrishnan, V. V., Brooks, M. S., Budge, S. M., & Dave, D., (2013). Fish processing wastes as a potential source of proteins, amino acids and oils: A critical review. *J. Microb. Biochem. Technol, 5*(4), 107–129.
20. Gildberg, A. R., (2004). Enzymes and bioactive peptides from fish waste related to fish silage, fish feed and fish sauce production. *Journal of Aquatic Food Product Technology, 13*(2), 3–11.
21. Gonçalves, J. M. S., Stergiou, K. I., Hernando-Casal, J. A., Puente, E., Moutopoulos, D. K., et al. (2007). Discards of trammel nets in southern European small-scale fisheries. *Fisheries Res, 8*(13), 5–14.
22. Green, J. H., Paskell, S. L., & Goldmintz, D., (1977). Fish peptones for microbial media developed from red hake and from a fishery by-product. *Journal of Food Protection, 40*(3), 181–186.
23. Hardy, R. W., Sealey, W. M., & Gatlin, D. M., (2005). Fisheries by-catch and by-product meals as protein sources for rainbow trout *Onchorynchus mykiss*. *J. World Aquacult. Soc, 36*(3), 393400.
24. Harnedy, P. A., & FitzGerald, R. J., (2012). Bioactive peptides from

marine processing waste and shellfish: A review. *J. Funct Foods, 4*, 6–24.

25. Hevia, P., Whitaker, J. R., & Olcott, H. S., (1976). Solubilization of a fish protein concentrate with proteolytic enzymes. *Journal of Agricultural and Food Chemistry*, *24*(2), 383–385.
26. Holub, D. J., & Holub, B. J., (2004). Omega-3 fatty acids from fish oils and cardiovascular disease. *Molecular and Cellular Biochemistry*, *263*(1), 217–225.
27. Hordur, G., K. & Rasco, B. A., (2000). Fish protein hydrolysates: Production, Biochemical, and functional properties. *Crit. Rev. Food Sci. Nutr, 40*, 43–81.
28. Jackson, A., (2009). The continuing demand for sustainable fishmeal and fish oil in aquaculture diets. *Int. Aquafeed, 12*, 32–36.
29. Kandra, P., Challa, M. M., & Jyothi, H. K. P., (2012). Efficient use of shrimp waste: Present and future trends. *Applied Microbiology and Biotechnology*, *93*(1), 17–29.
30. Khora, S. S., (2013). Marine fish-derived bioactive peptides and proteins for human therapeutics. *Int. J. Pharm. Pharmaceu. Sci, 5*(3), 31–37.
31. Kim, D. Y., & Lee, J. S., (2015). Directions for eco-friendly utilization and industrialization of fishery by-products. *Journal of Fisheries and Marine Sciences Education*, *27*(2), 566–575.
32. Kim, S. K., & Mendis, E., (2006). Bioactive compounds from marine processing by-products: A review. *Food Res. Int, 39*, 383–393.
33. Kim, S. K., & Wijesekaraa, I., (2010). Development and biological activities of marine-derived bioactive peptides: A review. *J. Func. Foods, 2*, 1–9.
34. Kotzamanis, Y. P., Alexis, M. N., Andriopoulou, A., Castritsi-Cathariou, I., & Fotis G., (2001). Utilization of waste material resulting from trout processing in gilthead bream (*Sparus aurata* L.) diets. *Aquacult. Res., 32*(1), 288–295.
35. Kumari, S., Rath, P., Kumar, A. S. H., & Tiwari, T. N., (2015). Extraction and characterization of chitin and chitosan from fishery waste by chemical method. *Environmental Technology and Innovation*, *3*, 77–85.
36. Madhavan, P., & Ramachandran, N. K. G., (1974). Utilization of prawn waste: Isolation of chitin and its conversion to chitosan. *Fishery*

*Technology*, *11*(1), 50–53.

37. Mahro, B., & Timm, M., (2007). Potential of biowaste from the food industry as a biomass resource. *Engin in Life Sci, 7*(5), 457−468.
38. Maralit, B. A., Aguila, R. D., Ventolero, M. F. H., Perez, S. K. L., Willette, D. A., & Santos, M. D., (2013). Detection of mislabeled commercial fishery by-products in the Philippines using DNA barcodes and its implications to food traceability and safety. *Food Control*, *33*(1), 119–125.
39. Mori, T. A., Vandongen, R., Beilin, L. J., Burke, V., Morris, J., & Ritchie, J., (1994). Effects of varying dietary fat, fish, and fish oils on blood lipids in a randomized controlled trial in men at risk of heart disease. *The American Journal of Clinical Nutrition*, *59*(5), 1060–1068.
40. Naylor, R. L., Hardy, R. W., Bureau, D. P., Chiu, A., Elliot, M., et al., (2009). Feeding aquaculture in an era of finite resources. *Proc. Natl. Acad Sci. USA, 106*, 15103–15110.
41. NOAA/USDA Alternative Feeds Initiative, (2011). The Future of Aqua Feeds (pp. 1–103). Technical Memorandum NMFS F/SPO-124.
42. Olsen, R. L., Toppe, J. E., & Karunasagar, I., (2014). Challenges and realistic opportunities in the use of by--products from processing of fish and shellfish. *Trends Food Sci. Technol., 36*, 144–151.
43. Péron, G., Mittaine, J. F., & Le Gallic, B., (2010). Where do fishmeal and fish oil products come from? An analysis of the conversion ratios in the global fishmeal industry. *Mar. Policy, 34*, 815820.
44. Rochet, M. J., & Trenkel, V. M., (2005). Factors for the variability of discards: Assumptions and field evidence. *Can J. Fish Aquat. Sci., 62*, 224–235.
45. Rustad, T., (2003). Utilization of marine by-products. *Electr. J. Environ. Agricult. and Food Chem., 2*(4), 458−463.
46. Shepherd, C. J., & Jackson, A. J., (2013). Global fishmeal and fish-oil supply: Inputs, outputs and markets. *J. Fish Biol., 83*, 10461066.
47. Shepherd, C. J., & Jackson, A. J., (2013). Global fishmeal and fish-oil supply: Inputs, outputs and marketsa. *Journal of Fish Biology*, *83*(4), 1046–1066.
48. Stone, F. E., & Hardy, R. W., (1986). Nutritional value of acid stabilized silage and liquefied fish protein. *Journal of the Science of Food and Agriculture*, *37*(8), 797–803.

49. Tacon, A. G. J., Hasan, M. R., & Metian, M., (2011). *Demand and Supply of Feed Ingredients for Farmed Fish and Crustaceans: Trends and Prospects* (pp. 80–87). FAO Fisheries and Aquaculture Technical Paper No. 564 FAO.
50. Tacon, A. G., & Metian, M., (2008). Global overview on the use of fish meal and fish oil in industrially compounded aqua feeds: Trends and future prospects. *Aquaculture*, *285*(1–4), 146–158.
51. Torstensen, B. E., Espe, M., Sanden, M., Stubhaug, I., Waagbø, R., Hemre, G. I., & Berntssen, M. H. G., (2008). Novel production of Atlantic salmon (Salmo salar) protein based on combined replacement of fish meal and fish oil with plant meal and vegetable oil blends. *Aquaculture*, *285*(1–4), 193–200.
52. Vezzulli, L., Chelossi, E., Riccardi, G., & Fabiano, M., (2002). Bacterial community structure and activity in fish farm sediments of the Ligurian Sea (western Mediterranean). *Aquacult. Int, 10*(2), 123–141.
53. Vezzulli, L., Moreno, M., Marin, V., Pezzati, E., Bartoli, M., & Fabiano, M., (2008). Organic waste impact of capture-based Atlantic bluefin tuna aquaculture at an exposed site in the Mediterranean Sea. *Estuar. Coast Shelf Sci., 78*(2), 369–384.
54. Virtanen, J. K., Voutilainen, S., Rissanen, T. H., Mursu, J., Tuomainen, T. P., Korhonen, M. J., & Salonen, J. T., (2005). Mercury, fish oils, and risk of acute coronary events and cardiovascular disease, coronary heart disease, and all-cause mortality in men in eastern Finland. *Arteriosclerosis, Thrombosis, and Vascular Biology*, *25*(1), 228–233.

# 8

# *Innovations in Production, Ecology, and Economy of Fisheries*

## CONTENTS

## 8.1. INTRODUCTION

Gelatin is a type of protein that is soluble in water, acquired economically by either pigskin or cowhide containing collagen. Collagens are transformed into gelatin by the process of hydrolysis. Gelatin is commonly utilized in the photography, foodstuff, medicinal industries, owing to the unusual characteristics of producing gel at normal temperatures. Many researchers have stated, 'substantial amount of gelatin might be manufactured by using fish bones and skins' (Jamilah and Harvinder, 2002; Batista et al., 2004) (Figure 8.1).

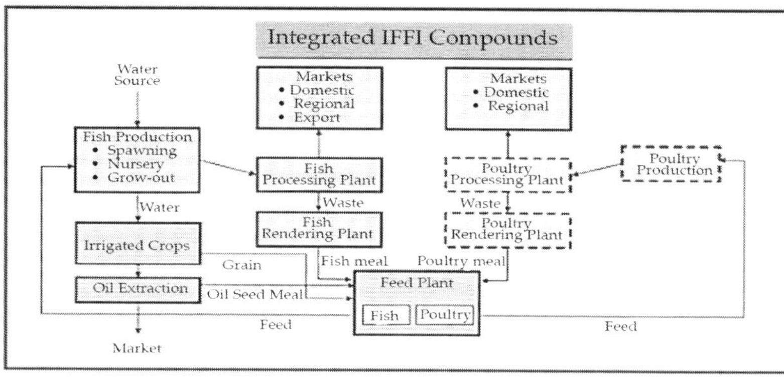


**Figure 8.1:** Innovation in fish and poultry production.

A big commodity of India's seafood exports is Surimi. By 2006, Indian surimi total output was thirty-five thousand tons (FAO, 2006). A significant amount of waste is produced throughout the processing of surimi. Fish solid wastes make up 50 to 70% of the initial raw resources, based on the carcass meat extraction process. Around 30% of these wastes comprise of heavy-collagen bone and skin. As raw resources for the profitable value-added commodity, like fish skin and bone gelatin, this waste has gained growing interest (Jongjareonark et al., 2005).

Gelatin is an essential biopolymer used in industrial gelling. It is used to improve aqueous system viscosity and shape aqueous gels. The useful characteristics include traditional rheology defined as melt-in-the-mouth, thermo-reversibility, and superb taste release (Jones, 1977). The current gelatin origins raise issues with religions like Islam and Judaism. Such cultures are reluctant to accept gelatin of pork, so beef gelatin is only permissible if it is manufactured on the base of religious criteria. The fish products having detachable scales are permissible with minor constraints in Judaism, although all types of fishes are permissible in Islam. Besides religious requirements, due to the large amounts of such materials produced, the commercial utilization of fish bones and skins, which are typically discarded, to produce extra revenue has waste disposal and economic gains for the fish industries. Gelatin is a multipurpose component utilized as a thickener, gelling agent, emulsifier, film former and stabilizer in foodstuffs, cosmetics, documentary films and pharmaceuticals (Rönnbäck, 1999). Like thermo reversible hydrocolloid, the difference among its gelling and melting temperatures, both temperatures are less than human body temperature. Gelatin is often extracted from the skins and bones of cattle, and fur of pigs. Many substitute materials have achieved interest from both industry and researchers not only because they address theological issues raised by Muslims and Jews and because they give technical advantages over the gelatins of mammal in certain situations. Fish skins from many types of fish are amongst the other materials that have been researched thoroughly as inputs for the production of gelatin. The skins of fish contain the considerable ability to manufacture high-quality gelatin with various gelling and melting temperatures with a broader range as compared to the gelatins of mammal (Campling et al., 2012).

The second natural widespread biopolymer is Chitin that is present in the insect's cuticles, the fungi cell walls, and the crustaceans shell (Knorr, 1984). Although chitin contains several useful biological characteristics including biodegradability, hemostatic action, wound healing, and biocompatibility characteristics and the considerable interest has also been given for its medicinal uses (Farkas, 1990; Fleet and Phaff, 1981). Because of its powerful hypocholesterolemic and lipid binding effects, but also as a film-forming agent, chitosan has been utilized as a supplement of food (Jeon, Arachchi, and Shahidi, 1999). Chitin, chitosan, and its compounds may be utilized in the industry of foodstuff as a stabilizing, emulsifying, antimicrobial, and thickening agent (Shahidi et al., 1999).

Chitin is mostly extracted from crustaceans and therefore, the sources of chitosan and chitin were the shells of prawn, shrimp, and crab. Shellfish production waste (for example, prawn, shrimp, lobster, crab, crawfish, and squid) produces about fourteen to 35% dry weight chitin, and such sources of chitin materials are becoming an increasing issue of waste disposal for the industries of fish and shellfish production. Japan is a top consumer and manufacturer of chitosan and chitin. Chitosan is a naturally occurring carbohydrate biopolymer produced from chitin deacetylation, the main component in crustacean shells like crawfish, shrimp, and crab. The second most widespread natural biopolymer is chitin after cellulose (Meyers and No, 1989). Chitosan is a polymer that is biocompatible, eco-friendly, and non-hazardous. From the past few years, chitosan is a polymer of chitinous, have gained great interest as among the promising sustainable polymeric materials for their widespread uses in the biomedical and pharmaceutical sectors for enzyme disinfection and immobility, in chemical wastewater processing plants and foodstuff industries for nutrition composition like gelling, binding, stabilizing, and thickening agent (Gordon, 1954; Knorr, 1984).

We easily obtain chitosan from different crustaceans shell waste such as shrimp, antarctic krill, prawn, crawfish, lobster, and squilla (Knorr, 1991). In Japan, the large amount of both chitosan and chitin are manufactured commercially (Brine and Austin 1981) although its quantity is less in India. The waste of shrimp shell is of considerable concern because it is India's largest single fish production waste. In India, the supply of shrimp production waste is calculated at 1, 35,409.5 MT per annum (Fisheries Statistics handbook, 2014). The shrimp waste or shrimp head approximately comprises about 50 to 60% of the shrimp bodyweight (Shanmugan et al., 2001). The output of aquaculture in India for prawn *Macrobrachium rosenbergii* from 2012 to 2013 is 3332 MT (Fisheries Statistics Handbook, 2014). That animal head weight is approximately 60% of the total weight of the body. Therefore, the supply of prawn waste maybe around, 1999.2 MT per annum.

The meal of fish is a versatile feed product containing good digestible nutrient, protein, and mineral content. Fish meal is used as a supplemental food to improve meat or egg production. Fish meal is deemed an essential poultry feed by combining it in the appropriate proportion in compound feeds. As the household and poultry animals diet are sometimes insufficient in vitamins and proteins, the meal of a fish is now very common because it offers essential nutrients and keeps it going (Vásquez, 2008).

The wastes of Fish generated from commercial fish processing activities are mostly comprised of flesh offal, hides, teeth, entrails, shells etc. Typically speaking, companies that prefer to remain safe efficiently will utilize waste from fish processing in the manufacturing of new products, not just contribute to the protection of the environment but also enhance their profits. Fish waste reflects the half volume of the total industry's raw material and is a cost-effective nutritional source (Oetterer, 2002). The perfect choice of solution is to use surplus material for making goods and prevents surplus production. It may be obtained by producing low-cost fisheries silage goods. The item has the reasonable nutritional quality and can be suitable for feeding livestock. The transformation of fish silage will efficiently turn fish waste into fish feed. This technique is also secure, cost-efficient, and environmentally sustainable. The silage of fish is a dark, sustainable liquid stock feeding produced by practical application of acids (acid fish silage) to acidify fish waste.

Preparation of fish silage generally relies on the raw resources and circumstances available locally (Hasan, 2003). Formic acid is the safest option for chemical silage preparation, the silages that are produced by formic acid are not extremely acidic and thus do not need neutralization before usage (Oetterer, 2002).

The key objective of this research was to extract Gelatin from the industry of surimi skin waste, to produce fish silage and fish meal from fish waste, to extract Chitin and Chitosan from shellfish waste. The physicochemical characteristics of the shellfish and fish produced through both by-products were studied. Production of different by-products from fish and shellfish waste is to improve coastal people's socioeconomic status by generating extra revenue. Therefore, chitosan, and chitin is an environmentally friendly commodity, it helps to preserve the biodiversity of the climate (Clark and Clausen, 2005).

## 8.2. CHARACTERISTICS OF FISH PRODUCTS

### 8.2.1. Raw Materials: Shrimp

Shell waste (*Parapenaeopsis stylifera*) was obtained from a nearby shrimp plant. On receiving, the waste from the shell was preserved at 20°C until used. Directly from the fishermen, the prawn (*Macrobrachium rosenbergii*) was gathered at Karla (Ratnagiri). As mentioned earlier, the Prawn shell waste was preserved. The shell waste was cured at 60°C for 24 hours in a warm air oven, chilled, and wrapped in polyethene containers, and then placed in a dried spot (Sukumaran, 1987). Dried shells were by hand pulverized. (*Scylla cerata*) Crab shells were gathered from the nearby market of fish. On receiving, the waste from the shell was preserved at 20°C until used. Fish wastes were picked from the market of fish and fish farming waste. On receiving, the fish waste preserves at 20°C used (Almeida and Vianna, 2005) (Figure 8.2).

**Figure 8.2:** Image showing shrimps before processing.

*Source: https://www.intrafish.com/finance/avanti-feeds-profit-dives-60-in-q2-on-less-farming-high-raw-material-costs/2-1-495379.*

### 8.2.2. Chemicals

Hydrochloric acid, calcium hydroxide, Acetic acid, sodium chloride, lactic acid, sulfuric acid, and sodium hydroxide were acquired from MERCK in India. All the utilized chemicals were of experimental grade (Meyer et al., 1979).

#### *8.2.2.1. Preparation of Chitin and Chitosan*

Vaporized and dry shells of prawn, crab, and shrimp were deproteinized at 1%, 2%, 3% sodium hydroxide solution at 85°C for thirty minutes in a water bath with continuous stirring at 1:15 (w/v) solid to solvent ratio. For the neutralization of pH, deproteinized shell wipes away with distilled water. Triplicate specimens were tested. This studied the content of proteins in processed shells. De-proteinized prawn and shrimp shell waste were de-mineralized in two ways; a. the waste of de-proteinized shell was processed in a water bath with 1 N hydrochloric acid for thirty and sixty minutes at 50°C; 120 minutes without heating; and at ambient temperature, with 2 N hydrochloric acids for 180 minutes. Shell waste was treated with distilled water after processing was done until the pH was balanced. The ash content of the de-mineralized shells was further studied. The de-mineralized shells were processed for discoloration utilizing various chemicals like industrial bleaching powder (2%, 3%, 4%, 5%, and 10%), KMnO4 (2%, 3%, 4%, and 5%), sodium hypochlorite (1%, 2%, 3%, and 4%) and $H_2O_2$ (3%, 6%, and 10%), at ambient temperature. A domestic kitchen grinder was used to ground the dry chitin, and rinsed with >1 mm blended sieves. This was bundled into a container made of polythene and kept at ambient temperature.

The procedure was performed in two forms for the deacetylation of chitin;

a) Specimen of Chitin was taken in a glass jar placed in a water bath with a 50% sodium hydroxide solid to solvent ratio 1:15 (w/v). The process was managed two hours, three hours, and four hours at 85°C.

b) The respective mixture was thoroughly sterilized for fifteen, thirty, and forty-five minutes for fifteen psi/121. The washing of the processed chitosan was done in tap water before pH 7.0.

#### *8.2.2.2. Preparation of Gelatin*

The method of Koli et al. (2011) was utilized to extract the gelatin from the skin of fish. The bones and skins of fish waste were utilized to extract the gelatin. Thawed bones and skins were properly rinsed and washed with extra tap water to eliminate superfluous content and processed for 40 min by soaking with sodium hydroxide solution of 0.2% (w/v). It was then saturated for 40 min with sulfuric acid of 0.2% (w/v). This was preceded by soaking with citric acid of 1.0% (w/v). The bones and skins were treated under flowing tap water after every soaking procedure, as they had a pH of around 7. For every forty minutes, the soaking and washing procedure was repeated for a complete-time of three hours. On each processing, the bones and skins ratio to the cleaning solvent utilized was 1:7 (w/v) of alkali or acid solution. Then the skins underwent final washing with purified water to eliminate any leftover content. The final extraction was conducted for seventeen hours in purified water at a stable temperature within the 55 ± 2°C range. The transparent extract collected was screened with Whatman filter paper in a Buchner funnel; the vacuum was dried, polished, and sealed in an airtight jar (Meyer and Schnick, 1978).

### 8.2.3. Preparation of Fish Meal

Wet rendering is used for the manufacturing of offal and fatty fish where sequential manufacturing of oil and fishmeal is proposed. According to the research project's goal, the fish meal was prepared by utilizing wet rendering. One kilogram of oil sardine taken in a steel vessel and then boiled for one hour on a gas stove with 1:5 (w/v) for preparing of fish meal. Once the vessel has cooled down and permitted to settle. By using a spoon, the oil which floats onto the surface is gathered. It is boiled again with intermittent stirring for another half hour and permitted to stand. The oil's upper layer is poured and stored separately. In a canvas bag, the cooked mass is taken and pressed in a screw press drained the oil surface layer and the fabric of cotton submersed the fish meat or residue (Bills et al., 1981). After this process, we stored it for 24 hours at a temperature of 45°C in an electric oven and after drying, after grinding in a mixture and at the end, we obtained the fish meal.

### 8.2.4. Preparation of Fish Silage

Fish silage is a protein produced from liquefied food. Fish silage may be manufactured from entire or portion of the fish by using organic acid (formic acid) or mineral acid (sulfuric acid) to process it. For the manufacturing of uniform sulfuric acid silage and formic acid silage, fish mashed meat was weighed and distributed into similar lots of 1 kg. According to Deepak et al. (2002), organic acid silage was produced by introducing 85% formic acid to modify pH.

### 8.2.5. The yield of Extracted Gelatin

The amount of acid was calculated. Sodium benzoate is then at a dose of One gm. And below 0.2 gm. BHT was applied to a broad-mouth plastic container in 1 kg of mashed fish meat. Such materials have been extensively blended and stored at room temperature. The mixture was frequently mixed to blend and to prevent mold production on the surface. The cycle lasted for twenty-one days.

#### *8.2.5.1. Biochemical Analyses*

The measurement of ash, fat, and moisture and pH values, and protein content, were calculated by using the AOAC (1990) method.

#### *8.2.5.2. Determination of Gel Strength*

The gelatin gel was formulated and the bloom value (strength of gel) of gelatin gel was measured using the process Wainewright (1977) describes. The gel was formulated in bloom container (SCHOTT GLAS, Mainz, Bloom test flask, Product No. 2112501) by mixing a dry gelatin powder of 6.67% (w/v) in purified water at 60°C. It was then chilled at ambient temperature for Fifteen minutes, and stored for development at 7°C for eighteen hours.

The strength of the gel was calculated on the British Standard BS 757 (BSI, 1975) TA. XT2i Surface Analyzer (Godalming, Scarsdale, N. Y./Stable microsystems, Surrey, Surface Technologies corp., U. K) with a load cell of five kilograms, cross-head speed of one millimeter per second and a flat bottom plunger of 0.5-inch diameter. The regular glass bloom container was precisely positioned under the plunger, and then the penetration examination was carried out. The highest possible force was calculated (in gram) before the probe entered the gel to a depths of four millimeters.Readings were made in triplicate utilizing identical glasses, and the maximal penetration intensity was measured and is used to measure the strength of the gel.

#### *8.2.5.3. Hydroxy Proline Content of Gelatin*

The gelatin hydroxyproline content was calculated with minor adjustment as per the Loxley and Bergman methods (1963). The specimens were hydrolyzed in a reflex condenser with 6 M Hydrochloric acid at 1100°C for 24 hours and filtrated through filter paper Whatman No.1. The filtrate was neutralized to pH 6.0 to 6.5 with 1 M sodium hydroxide. The neutralized sample (0.1) milliliter was moved into a test tube and well applied and combined isopropanol (0.2 ml). 0.1 milliliter of an oxidant solution (a combination of 7% (w/v) of chloramine T and acetate/citrate buffer, pH 6, was applied and completely blended at a ratio of 1:4 (v/v). After which 1.3 mm of Ehrlich's reagent solution (a solution mixture two gram of benzaldehyde p-dimethyl-amine in 3 mm of isopropanol) was applied. The mixture was blended and heated in a water bath at 60°C for 25 minutes after which refrigerated in potable water for two to three minutes. The solution was isopropanol-diluted to 5 mm. The absorbance was determined with a spectrophotometer (Thermo Spectronic, UV 0628) toward the water at 558 nanometers. Already tested at the same period were normal hydroxyproline solutions with concentrations varying from ten to sixty ppm. The content of hydroxyproline was measured and expressed in mg/g sample shape.

### 8.2.6. Determination of Viscosity Gelatin

Solutions of gelatin at a concentration of 6.67% (w/v) is formulated for viscosity calculation by dissolving the arid powder in sterile water and heating at 60°C. The solution's viscosity (cP) of 10 ml was measured by utilizing a Brookfield Digital Viscometer (Model DV-E Brookfield Engineering, United States of America) fitted with a 40°C ±1°C No.1 spindle (Cho et al., 2006).

### 8.2.7. Emulsifying Capacity and Stability

Yasumatsu et al. (1972) approach were used to assess the efficiency and consistency of the emulsification. We formulated the emulsion with one gram of each specimen, 50 mm of cold sterile water (4°C) and 50 mm of sunflower oil. The gelatin specimens were spread using a homogenizer. Every blended sample was in 50 mm centrifuge tubes in equal measure. For ten minutes, one centrifuge tube was centrifuged immediately at 4000×g and the other centrifuged under the same circumstances following thirty minutes of heating in a water bath at 80°C and cool at room temperature (25°C).

### 8.2.8. Determination of Melting Point of Gelatin

Measurement of the melting point was achieved using an updated procedure from Wainwright (1977). Solutions of gelatin were formulated at 6.67% (w/v) and 5 mm aliquot of each specimen was moved to a narrow glass tank (borosilicate tank, 12 mm, 75 mm). The specimens had been thoroughly rinsed for five minutes in vacuum desiccators. The tubes were filled with Para film and heated for fifteen minutes in a 60°C water bowl. The tubes were subsequently refrigerated in ice-cold water and stabilized at 10°C. For 18 hours five drops were put on the gel surface of a combination of 75% chloroform and 25% reddish-brown dye (food color). The gels were placed in a 10°C water bath and heated the bath at a pace of 0.2°C to 0.4°C per minute. The temperature of the bath was read using an electronic digital thermometer (Fisher Scientific). The bath temperature was analyzed using a modern, automated (Fisher Scientific) thermometer. The temperature was calculated as the melting point in which the dye drops started to travel freely down the gel.

### 8.2.9. Water Holding Capacity

Water holding capacity (WHC) was calculated utilizing the centrifugation process (Martin and Diniz, 1997). Duplicate specimens (0.5 g) of gelatin in 20 mm of water is dissolved and spread in centrifuge tubes, with a vacuum blender for the thirty seconds. The dispersion was permitted to remain for Six hours at ambient temperature, and thirty minutes at 2800×g at centrifuge. The centrifugation was extracted with filter paper Whatman No.1 and the amount retrieved was calculated. It was measured the gap among the original amount of sterile water applied to the protein sample and the supernatant quantity, and the effect was recorded as milliliters of water consumed per gram of gelatin specimen.

## 8.2.10. Experimental Results

Table 8.1 indicates the closely related components of the materials utilized to remove gelatin, while Table 8.2 displays the closely related composition of processed gelatin (Koli et al., 2015)

**Table 8.1:** Proximate Compositions of Different Fish Skin Used for Extraction of Gelatin

| Sr. No. | Species Name/Source of Raw Materials | Fat (%) | Moisture (%) | Ash (%) | Proteins (%) |
|---|---|---|---|---|---|
| 1. | Pink Perch –Rani Masa (*Nemipterus japonicus*) | 1.41 ± 0.11 | 79.63 ± 2.07 | 1.29 ± 0.06 | 16.76 ± 0.30 |
| 2. | Tiger toothed croaker (*Otolithes rubber*) Dhoma | 1.51 ± 0.20 | 79.50 ± 2.84 | 1.78 ± 0.79 | 17.50 ± 0.81 |
| 3. | Tilapia (*Oreochromis mossambicus*) | 2.20 ± 0.26 | 68.63 ± 2.11 | 1.60 ± 0.17 | 25.07 ± 0.19 |
| 4. | Black King Fish – Modusa (*Ranchycentron cana-dus*) | 12.00 ± 0.28 | 70.00 ± 0.98 | 4.50 ± 0.01 | 12.94 ± 0.44 |
| 5. | Ghol (*Protonibea dicanthus*) | 2.48 ± 0.29 | 75.80 ± 1.249 | 1.06 ± 0.05 | 20.63 ± 0.55 |
| 6. | Horse mackerel (*Mega-laspis cordyla*) | 11.50 ± 20 | 76.80 ± 3.1 | 2.75 ± 1.6 | 17.90 ± 0.18 |
| 7 | Shrimp (*Parapenaeop-sis stylifera*) | 0.17 ± 0.4 | 76.06 ± 1.63 | 6.14 ± 0.07 | 4.73 ± 0.09 |
| 8. | Catla (*Catla Catla*) | 3.22 ± 0.04 | 74.77 ± 0.40 | 3.51 ± 0.08 | 18.87 ± 0.05 |
| 9. | Crab (*Scylla serrata*) | 0.20 ± 0.50 | 28.71± 1.63 | 7.01 ± 0.30 | 3.80 ± 0.27 |
| 10. | Prawn (*Macrobrachium rosenbergii*) | 0.27 ± 0.5 | 72.29 ± 1.1 | 8.01 ± 0.37 | 2.85 ± 0.17 |

**Table 8.2**: Proximate Composition of Extracted Gelatin from Different Fish Skin

| Sr. No. | Species name/Source of raw materials | Fat (%) | Moisture (%) | Ash (%) | Proteins (%) |
|---|---|---|---|---|---|
| 1. | Pink Perch – Rani Masa (*Nemipterus japonicus*) | 0.55 ± 0.10 | 8.73 ± 1.20 | 0.30 ± 0.01 | 72.63 ± 1.01 |

| | | | | | |
|---|---|---|---|---|---|
| 2. | Tiger toothed croaker-Dhoma (*Otolithes rubber*) | 0.85 ± 0.10 | 9.60 ± 0.72 | 1.88 ± 0.14 | 86.45 ± 0.53 |
| 3. | Tilapia (*Oreochromis mossambicus*) | 1.27 ± 0.08 | 9.53 ± 0.23 | 0.94 ± 0.05 | 88.29 ± 0.47 |
| 4. | Blacking fish – Medusa (*Ranchycentron canadus*) | 0.47 ± 0.02 | 6.04 ± 0.07 | 2.24 ± 0.14 | 88.72 ± 0.41 |
| 5. | Ghol (*Protonibea dian-thus*) | 0.27 ±0.067 | 8.43 ± 0.86 | 0.93 ±0.077 | 90.36 ± 0.47 |
| 6. | Catla (*Catla Catla*) | 0.86 ± 0.03 | 8.13 ± 0.07 | 1.53 ± 0.06 | 89.45 ± 0.36 |
| 7. | Horse mackerel (*Megalaspis cordyla*) | 1.5 ± 0.70 | 6.12 ± 0.00 | 3.01 ± 0.10 | 91.07 ± 0.10 |

The strength of the gel of extracted gelatin was depended primarily on the molecular weight, the extraction temperature and acid-alkali concentrations (Hansen and Ockerman, 1988). The strength of the gel is one of the most significant physical characteristics of gelatin and generally, the gelatin of fish has weaker strength of the gel than the gelatin of mammalian.

The strength of the gel of commercial gelatin represented as bloom value varies from 100–300 g, however, most desirable is gelatin with bloom values within the range of 250–260 g (Holzer, 1996). The present analysis results indicate that in table 8.3, the skins of various organisms produced gelatines with specific bloom values. The current study's on the bloom value was reported as 140–422 g.

The gelatin melting point derived from tilapia is higher than that of the melting point of the gelatin derived from other organisms examined in table 8.3. Such melting points are much higher than those recorded in the 8 to 10°C range for cold skin. The gelatin of the fish is therefore considered to have a low melting point than the gelatin of mammalian (Norland, 1990). Gelatin's melting point increases with the rise of its molecular weight (Courts and Ward, 1977). Regenstein and Choi (2000) stated that with the increase of maturation temperature the melting point also increases. The structure of amino acids can also contribute to the properties of a melting point (Norland, 1990).The current study shows that the gelatin viscosity derived from various fish skins was observed to be within the range of 6.90–85.53cP (Table 8.3). The second most significant industrial feature of gelatin after the strength of the gel is viscosity (Courts and Ward, 1977). Viscosity is partly regulated by the molecular weight, it is a distribution of molecular size (Sperling, 1985). Most of the commercial gelatins have recorded viscosities of up to 13.0 cP (Johnston-Banks, 1990). Harvinder and Jamilah (2002) recorded that the viscosity of black tilapia gelatin and red tilapia gelatin was 7.12 cP and 3.20 cP respectively, for grass carp, rohu, and common carp it was 7.07 cP, 6.06 cP and 5.96cP respectively (Ninan et al., 2011) and while it was 3.23 cP for channel catfish (Yang et al., 2007), 3.40 cP and 5.24 cP for snakehead and catfish respectively (See et al., 2010). The pH variations are known to affect the viscosity and there was a minimal gelatin viscosity in the range of 6 to 8 (Stainsby, 1987).

**Table 8.3:** Rheological Properties of Gelatin Extracted from Different Fish Skin

| Sr. No. | Species name/Source of raw materials | Melting Point (°C) | Viscosity (cp) | Bloom value (g) |
|---|---|---|---|---|
| 1. | Pink Perch-Rani Masa (*Nemipterus japonicus*) | 19.23 ± 0.26 | 8.47 ± 0.60 | 140 ± 2.88 |
| 2. | Tiger toothed croaker (*Otolithes rubber*) Dhoma | 20.36 ± 0.29 | 10.53 ± 0.46 | 170 ± 2.88 |
| 3. | Tilapia (*Oreochromis mossambicus*) | 27.48 ± 0.05 | 6.90 ± 0.06 | 190.24 ± 3.24 |
| 4. | Black King Fish-Modusa (*Ranchycentron canadus*) | 22.13 ± 0.12 | 13.53 ± 0.57 | 222 ± 1.15 |
| 5. | Catla (*Catla catla*) | 24.50 ± 0.20 | 9.50 ± 0.20 | 180.76 ± 0.28 |
| 6. | Ghol (*Protonibea dicanthus*) | 23.433 ± 3.22 | 13.76 ± 2.44 | 422.69 ± 4.22 |
| 7. | Horse mackerel (*Megalaspis cordyla*) | 21.13 ± 0.12 | 53.85± 0.24 | 136± 4.24 |

The current study shows that the yield of the probable skin gelatin composition ranges from 11.48% to 20.40% as seen in table 8.4, and the sum of soluble. The components of extraction yield in the skins as that of gelatin from skins varying from materials differ with the species and between 5.5% to 21% of the fish starting age. Also, the raw sources weight (Grossman, Jamilah, and Bergman, 1992; the difference in the method of extraction may also affect the yields. Harvinder, 2002; Muyonga et al., 2000;

Reporting the yield of gelatin, like dry gelatin weight Gimenez et al., 2005a; Gimenez et al., 2005b; and Songchotikunpan et al., 2008; compared to the wet skin weight is common, typical but not quite reliable). The variance in these values relies both on the Hafsteinsson and Gudmundsson variations (Muyonga et al., 2004).

**Table 8.4:** The Hydroxy-Proline Content Yield of Gelatin and pH Extracted from Different Fish Skin

| Sr. No. | Species name/Source of raw materials | Yield (%) | pH of 1% solution | Hydroxy proline (mg/g) |
|---|---|---|---|---|
| 1. | Pink Perch-Rani Masa (*Nemipterus japonicus*) | 11.48±2.4 | 4.60 ± 0.00 | 7.63 ± 0.02 |
| 2. | Tiger toothed croaker (*Otolithes rubber*) Doma | 12.50±3.2 | 4.62 ± 0.02 | 7.77 ± 0.02 |
| 3. | Tilapia (*Oreochromis mossambicus*) | 14.07±3.6 | 5.24 ± 0.02 | 7.88 ± 0.03 |
| 4. | Catla (*Catla Catla*) | 14.50±4.2 | 4.23 ± 0.15 | 8.87 ± 0.06 |
| 5. | Black King Fish - Modusa (*Ranchycentron Canadus*) | 20.40 ±3.5 | 4.81 ± 0.02 | 8.34 ± 0.43 |
| 6. | Horse mackerel (*Megalaspis cordyla*) | 12.03±4.2 | 5.30 ± 0.05 | 7.36 ± 0.00 |
| 7. | Ghol (*Protonibea dicanthus*) | 18.68±2.2 | 4.76 ± 0.03 | 7.70 ± 0.04 |

Gomez-Guillen et al. (2002) recorded a small difference in the yield of extractions among the species of fish (megrim: 7.4%; sole: 8.3%; hake: 6.5%; cod: 7.2%).

It's also clearly known that the bonds of hydrogen among free amino acid groups and water molecules in gelatin are important to the bloom value of gelatins (Arnesen and Goldberg. 2002). The greater the rates of hydroxyproline close to imino acid, the stronger the gelatin gel strength (Sarabia et al., 2000). The hydroxyproline amount of the skin of bovine and pork gelatin was 8.3%

(Gullen-Gomez, 2009), and 9.1% (Bhat and Karim, 2009) respectively. Throughout this analysis, as seen in table 8.4, the hydroxyproline amount of gelatin derived by chosen and optimized process ranged between 7.36 and 8.34%. This explains that the gelatin derived from various fish skins has reduced gel strength compared with the gelatin of mammalian (Ellis et al., 2014; García et al., 2008).In the current analysis, emulsifying stability and capacity of various fish skins gelatin were seen in Table 8.5. Emulsifiers are pavement-active components that accumulate to the interface and promote the creation of tiny droplets by reducing the interface while homogenization (Walstra, 2003). The amphoteric existence of the peptide chain with hydrophobic zones enables gelatin function like an emulsifier and is utilized in the manufacturing of water-in-oil emulsions and toffees, such as whipped cream, salad dressing, and low-fat margarine (Peterman and McAllister, 1992; He and Baziwane, 2003). PUFA is the operational characteristics of proteins in a system of food rely in part on the WHC that relates to proteins' abilities to drink water and maintain it against a gravitational force inside the matrix of protein. As seen in Table no. 8.5, the WHC of various fish gelatin was removed. The WHC of centrifuged gelatin renders it a good medium for which the loss of drip and impairing the juiciness in frozen meat or fish products when dried or fried, and when the hydrolyzed protein has experienced a significant loss of its ability to retain water. Koli et al. (2011) stated that Tiger-toothed croaker bone and skin gelatins had a WHC of 3.00 ml/g and 4.50 ml/g, while bone and Pink peach gelatin were 1.50 ml/g and 2.36 ml/g (Gowing and Momot, 1977).

**Table 8.5:** Functional Properties of Gelatin Extracted from Different Fish Skin

| **Sr. No.** | **Species name/ Source of raw materials** | **Water Holding capacity (ml/g)** | **Emulsifying stability (%)** | **Emulsifying capacity (%)** | **Foaming stability (%)** | **Foaming capacity (%)** |
|---|---|---|---|---|---|---|
| 1. | Pink Perch- Rani Masa (*Nemipterus japonicus*) | 2.36 ± 0.06 | 32.40 ± 0.60 | 47.50 ± 0.51 | 10.00 ±0.51 | 15.00 ±0.76 |
| 2. | Tiger toothed croaker (*Otolithes rubber*) Dhoma | 4.90 ± 0.56 | 35.70 ± 1.87 | 55.70 ± 0.76 | 14.00 ±1.96 | 17.00 ±1.60 |
| 3. | Tilapia (*Oreochromis mossambicus*) | 2.46 ± 0.06 | 51.53 ± 0.08 | 53.45 ± 0.06 | 12.00 ±0.50 | 14.00 ±0.76 |

| | | | | | | |
|---|---|---|---|---|---|---|
| 4. | Horse mackerel (*Megalaspis cordyla*) | 2.56 ± 0.09 | 51.53 ± 0.08 | 53.45 ± 0.06 | 13.00 ±0.50 | 16.00 ±0.70 |
| 5. | Black King Fish-Modusa (*Ranchycen-tron Canadus*) | 4.43 ± 0.08 | 32.50 ± 0.73 | 55.66 ± 0.48 | 12.00 ±0.20 | 12.00 ±0.60 |
| 6. | Catla (*Catla catla*) | 2.86 ± 0.08 | 52.70 ± 0.26 | 54.80 ± 0.26 | 12.00 ±0.50 | 18.00 ±0.70 |
| 7. | Ghol (*Protoni-bea dicanthus*) | 2.46 ± 0.06 | 55.53 ± 0.06 | 58.40 ± 0.06 | 14.00 ±0.50 | 16.00 ±0.86 |

The close composition of the crab, prawn, and shrimp shell wastes is seen in table 8.6. Table 8.7 shows the close concentrations of chitosan, chitin, fish silage, and fish meal.

The current study shows that the probable composition of crab shell, shrimp, and prawn waste was evaluated on the base of wet weight (Table -6). The value of ash, moisture, fat, protein, calcium carbonate and chitin were 7.01, 68.29, 0.17, 3.80, 14.7 and 4.27% for crab shell, 6.14, 76.06, 0.17, 4.73, 8.36% for shrimp and 8.01, 72.29, 0.27, 2.85, 11.7 and 5.72% for prawn waste respectively. Shrimp has higher protein and moisture percentage was found than the crab shell and prawn waste. Other probable composition specifically fat, chitin, ash was higher in prawn shell waste relative to crab shell and shrimp waste. Calcium carbonate was high in crab shell waste and a bit low in shrimp waste than the waste of prawn shell. Madhavan and Thankappan (1985) recorded on a dry weight basis the probable composition of the shell mixture in shrimp, except for moisture (percentage). They noted that ash, moisture, fat, and protein were in the range of 20.00%, 78.82%, 2.56% and 37.00% in *P. indicus*, 27.92%, 82.25%, 7.39% and 28.38% in *P. stylifera*, 23.15%, 78.15%, 3.36% and 34.04% in *M. Monoceros,* and 18.91%, 74.86%, 2.95% and 39.62% in *M. dobsoni,* respectively.

Bedekar and Pradhan (2002) stated on a dry weight basis a probable composition range for *P. monodon and P. indicus*. They acquired ash (in percentage), moisture (in percentage), acid-soluble protein (in percentage), calcium carbonate (in percentage), alkali-soluble protein (in percentage), chitin (in percentage), and dead pigments (in percentage)in the range of 16.9

to 17.2, 8 to 10, 13.0 to 13.8, 30.1 to 30.7, 26.8 to 27.6, 17.3 to 17.9 and 3 to 5 in *P. monodon* and 19.4 to 19.6, 8 to 10, 10.6 to 11.1, 34.6 to 35.0, 24.9 to 25.4, 12.1 to 12.5 and 3 to 5 in *P. indicus,* respectively. In the current study, the earlier researchers stated as probable composition value with findings made for crab shell, shrimp, and prawn waste (Sundström, 2012).High-fat fish such as herring, sardines, anchovies, etc. are commonly utilized as raw resources to generate fish meal in the current study. Limited by catch fish from shrimp trawling may usually not be sold as fresh fish due to various factors such as very limited scale, and may also be utilized as bony type. Filleting facilities waste and fish processing waste, other food wastes and cannery waste are frequently utilized as raw resources for fishmeal development. Probable yield and composition of fish silage and fish meal were shown in Table 8.7.

**Table 8.6:** Proximate Compositions of Shell Waste of Shrimp, Prawn, and Crab

| Sample | Calcium carbonate (%) | Ash (%) | Moisture (%) | Protein (%) | Chitin (%) | Fat (%) |
|---|---|---|---|---|---|---|
| Crab | 14.7 ± 0.56 | 7.01 ± 0.30 | 68.29 ± 1.1 | 3.80 ± 0.10 | 4.72 ± 0.27 | 0.17 ± 0.40 |
| Shrimp | 8.36 ± 0.41 | 6.14 ± 0.07 | 76.06 ± 1.63 | 4.73 ± 0.09 | 4.74 ± 0.27 | 0.17 ± 0.10 |
| Prawn | 11.7 ± 0.56 | 8.01 ± 0.37 | 72.29 ± 1.1 | 2.85 ± 0.175 | 5.72 ± 0.27 | 0.27 ± 0.50 |

Fish silage is a protein produced from liquefied fish. An option to preparing a fish meal is the storage of extra fish raw meat and fish as silage to be used in livestock feeding. Fish ensilage does not need facilities or significant capital, large setup and large machinery. The method is quite quick and suitable for operating even in distant fishing communities where the facilities are minimal for some form of production. Additional work is required, research work is underway to prepare the fish meal, silage, chitosan, and chitin while using various methods and various shell and fish waste and their relative usage in various fields.

**Table 8.7:** Proximate Compositions and Yield of Chitin, Chitosan, Fish Meal, and Fish Silage

| Sr. No | By-Products | Proximate composition (%) | | | | Yield (%) |
|---|---|---|---|---|---|---|
| | | Fat | Protein | Ash | Moisture | |
| 1 | Chitosan | 0.05± 0.5 | 2.29± 1.2 | 36.82± 1.6 | 4.5± 2.5 | 63.45±0.98 |

| | | | | | | |
|---|---|---|---|---|---|---|
| 2 | Chitin | 0.63± 1.8 | 4.93± 2.5 | 26.45± 1.5 | 12.90± 1.9 | 22.40±1.2 |
| 3 | Fish Silage | 2.13±0.14 | 15.78 ± 0.44 | 2.32 ± 0.17 | 79.47±0.42 | 65.70±0.25 |
| 4 | Fish Meal | 3.50 ± 1.2 | 71.80 ± 1.6 | 4.56 ± 1.40 | 7.50 ± 1.30 | 70.50±1.26 |

## 8.3. EXPLOITATION AND INNOVATIVE IN FISH-BY-PRODUCTS

The issue of the fair and efficient usage of raw resources appears to be extremely relevant to the worldwide seafood production industry. The quest for innovative technological and operational solutions is to make it easier to utilize the techniques which are of no utility and diminished commercial value (Bruno et al., 2019).Demand for new fish and fish-based goods continues a growing trend in fish production, producing substantial numbers of wastes and by-products along with vast quantities of minimal-value by-catch of fish that are not used effectively to optimize its economic capacity. More than 21 million tons of fish were killed or used for non-food purposes in 2012 (FAO, 2014) due to an assortment of unacceptable properties, such as significant heterogeneity, flavored, small size, unhealthy shape, bony structure, considerable fat content, the danger of toxic substances. As a consequence, just 50 to 60% of the by-catch of fish is required for human usage and 25% is dumped completely as waste (Rustad, 2003) (Figure 8.3).

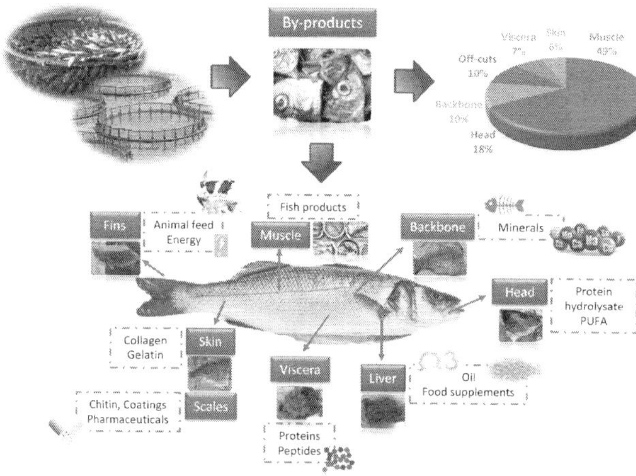


**Figure 8.3:** Modern trends and innovations in processing of fish by-products.

*Source: https://www.mdpi.com/1660-3397/17/12/689/htm.*

Research reveals that up to 50 to 80% of the raw fish sources may be discarded after processing as toxic waste, i.e., conventional round filleting. Botta and Shahidi (2012) list some significant fish processing by-products and rejects as a usable cheek, tongue, liver, and stomach, fish skin, residual fish proteins, carotenoid pigments, chitinous materials, gut enzymes, flavorings, and cold-temperature fish blood anti-freeze proteins that could be required for more processing. In Latvia, low-value by-products of fish farming and fish are sold as a raw resource with no additional benefit, refined into fishmeal or considered for standard output (for example, smoked sprats), therefore deciding substantial economic benefits for the whole sector (Mulye et al., 2014). Taking advantage of such varieties of fish and fish production by-products to produce new goods has tremendous potential for increasing value, reducing waste and encouraging their effective usage (Chalamaiah et al., 2012).

By-products of fish processing may be converted into different categories of goods having a higher market value such as pharmaceutical, food, and biotechnology devices. The consumer market has used to provide revolutionary and creative goods daily, which encourages businesses to create different product lines and utilize improved manufacturing technology. In the food field, the current study in this region is focused mainly at creating new methods for various forms of fish-based ground meat and producing a broad variety of items made on their bases, like sausage and culinary goods, analogues of delicacy items, canned, and preserved goods (Kalpana, Devi, and Aparna, 2013).

The key distinction between minced meat goods and entire fish-based items or their particular component (for example, fillet) is the degree in which the raw sources are utilized. Generally, only 30 to 40% of the fish in the classical fish sector is processed into high-quality items (medallions, fish steaks, or fillets) and 60 to 70% are by-products of fish processing have poor quality and limited economic value that are practically not in demand on the food sector. Also, the use of these surpluses in production is typically risky and generates substantial environmental and economic costs. However, fish processors are based on gaining the most profit from primary materials, with very little interest in by-products resulting in minimum opportunities of economic due to inadequate technological innovation that allows for these materials to be utilized (Geelen and Xu, 2018).

This segment intends to investigate the utilization possibilities of the by-products of fish processing and create new products for consumption with sufficient organoleptic quality that is suitable for diversification and development. The research also intended to recognize and resolve the technical difficulties of using sprat by-products as a raw source for manufacturing. Fish-based pâté portion (spreadable product) was created from by-products of sprat manufacturing, while fish-based terrine (typically served in slices) is in its developmental stage, as a result of the project. To broaden the variety, different variations (colors, additives, flavors,) have been generated from every basic product (Faasili and King, 1999).

### 8.3.1. Materials and Processes

Their research is based on the Baltic sprats, a tiny herring-like, aquatic fish found in the waters of the Baltic Sea and quite common to the regional economy of Latvia. The pâté was produced by using the fish processing waste, such as the heads and, in certain situations, the tails of smoked sprats (from an in-house laboratory, Piejūra Ltd., Nīca, Latvia). The study was carried out in 7 months, from Sept. 2017 to Mar. 2018. The present research has contributed to a further sequence of studies on the diversification of the product portfolio (Martin 1999).

The new 2015 data available at the United Nations Global Capture Production Food and Agriculture Organization database indicates that marine capture production for Sprattus has risen significantly from 2014 to 2015 by 36.9% and reached 677,029 in 2015 (FAO, 2017). While being an important part of the Latvian and local fisheries market, Baltic sprats are deemed an industrial species because the majority of the catch is produced into goods not expected for human use, primarily fishmeal (Chapman and Miles, 2015). In Latvia, only about 25% of captured sprats are transformed into products for human use, often into bottled sprats, typically smoked ("Sprats in Oil"). Therefore, the processing of canned sprat is very resource-consuming and does not produce substantial income. Additionally, although being a typical commodity in the area, the "Sprats in Oil" market is very small. Despite the vast geography of exports, the drug is used mostly by the local community and comparatively limited ex-pat populations abroad in Eastern Europe (Crook and Zint, 1998).

The research investigated the possibilities for technological advancement aiming at ready-to-eat food production from sprat processing by-products left until smoked fillets are further transformed into conventional "sprats of oil." The past study of fish processing by-products and the examination of the technical ability of pâté development from the by-products of sprat processing investigate the possibilities of novel fish-based commodity utilization in the food industry (González and Genovese, 1998; Sneddon, 2007).

### 8.3.2. Fish Processing By-Products and Their Economic Potential

There is currently an identified scarcity of good quality protein of animal worldwide (Pimentel and Pimentel, 2003). In developed countries, the key source of these husbandry and proteins livestock is reaching the surplus limits and it is hard to forecast several rises in production, despite the increasing demands of urbanisation and the huge growth of population. Simultaneously, there are fantastic growth opportunities for fish farming and commercial fishing. Lots of studies reveal that sprats are a good source of lipids, protein, vitamins, and nutrients, although having a limited amount of carbohydrates (Jayathilakan et al., 2012). Nevertheless, it should be noted that the variety of fish products is very small, and the industry faces the problem to create and introduce new fish products into the market, increasing the choice from the existing canned fish, raw, frozen varieties. Towards the other side, the processing of the fish product creates a considerable amount of by-products and contaminants with a strong nutrients content that is rejected for human use, such as ridges of fish heads and meat cuts. Generally, four major routes for fish processing by-products and waste disposal may be differentiated as per the possible value-added profitability and volume: aquaculture feed production, fertilizers, food development planned for human use, domestic, and livestock animals, and innovative solutions for biotechnology, medicinal, and related industries (Wilson et al., 2011; Sanmartín et al., 2012).

1. Fish-based products intended for human consumption make up the two areas of fish processing and provide considerably improved profitability, innovative solutions and food production like bioactive products for the biotechnology and pharmaceutical industries, along with the cosmetic industry. For example, fish-based collagen has been far less analyzed relative to mammalian collagen, but around 30% of fish processing waste is bone and skin with a high content of collagen. Nevertheless, whereas being the high-end solution sector, the largest-potential profitability sector is the most innovative and therefore needs reliable and substantial investment and technical capability. Innovative approaches for fish processing by-product extraction for the food

industry, while substantial innovation and information input are still needed, are more possible for scaling and application to our regional fish processing market.

2. Fish-based fertilizers have been utilized for decades in agriculture. The solution may not even need processing if, for example, rotten fish are discarded to the fields to maximize the vitality of the land. Fertilizer production from dead fish was introduced as early as the beginnings of the twentieth century but this approach has a very small profit. Regarding agriculture uses, nearly 10% of fish farming waste and by-products are utilized as organic fertilizers.
3. The next section is the processing of fishmeal's for aquaculture and domestic animals, primarily in the forms of fishmeal. Up to 30% of all fish and shellfish captures worldwide are processed into fishmeal and oil (Bimbo and Crowther, 1992). Meals and Chapman (2015) suggests that about 90% of fish meal is used in fish and animal diets, particularly pets, supplying about 14% of the world's animal protein requirement (Venugopal, Shahidi, and Lee, 1995). In Latvia, a product with considerably lower profit than food items designed for human use or innovative technical solutions is used to fulfill fishmeal production at 75% with sprats capture.

Rustad (2003) notes that consumer demand is an important factor fostering investment and innovation fueling increased usage of by-products for fish processing. In the increasing population, there is embedded in the consumer potential of advanced fish-related human consumption goods, even those focused on industrial organisms. Macnaughton, Parfitt, and Barthel (2010), for example, remember that by 2050 "prospects for feeding a population of nine billion" have to be sought. Furthermore, projections by the United Nations (2008) suggest that 70% of the world would reside in metropolitan areas by 2050, signaling a rapid shrinkage of the agricultural sector. Simultaneously, evidence reveals that up to 50% of the food is not coming on the market (de Fraiture, Lundqvist, and Molden, 2008). Most of the waste occurs in the period of post-harvest, although losses of the process may be reduced by incorporating more innovative methods and technology for processing (Jaczynski and Taskaya, 2009; Gehring et al., 2011).

Together with the changing customer preferences and accelerated speed of everyday life, these trends have also shown the tremendous demand opportunity for ready-made consumer goods. Additionally, the demand for ready-made fish-based food goods is projected to increase. As Barthel, Parfitt, and Macnaughton (2010) has shown, the rising demand for seafood-based goods is closely correlated with Bennet's law, suggesting a decrease in starchy food staples intake and a dietary shift to meat, dairy, fish, and vegetables and fruit as income rises. It would also raise the value-added that is generated by competition and only happens because the customer needs the commodity and can buy it (Salin, Atkins, and Salame, 2003). Additionally, It should be remembered that there is a growing competition on the market to ensure nutritional products are still available for low-income classes.

Therefore, a by-product of sprat processing more use in the manufacture of ready food products for human consumption may be known as the best solution presenting the maximum possible amount of value-added and profitability, although requiring relatively moderate spending. Simultaneously, it tackles both the urgent need for new food supplies and growing market demand in fish goods and the demand for low-cost segmented items. It provides an incremental source of income to the manufacturers while reducing waste management costs. Nevertheless, many technological challenges might be addressed in this production together with the creation of by-product of sprat-based food product with acceptable organoleptic characteristics, such that final consumers will be involved in purchasing it (Ramírez-Ramírez and Shirai, 2011).

### 8.3.3. Innovative Pâté Made of Smoked Sprats Processing By-Products Ready for Human Consumption

Throughout the food sector, the amount of ready-to-eat products that do not need culinary preparation and contain heterogeneous components is growing, reflecting the changing customer preferences. Types of these items present on the market include whey and fruit beverages, ham slice cheeses, hams, and cheese sausages etc. Although the phenomenon is not so evident in the fishing sector, the amount of ready-to-eat products is limited, and the condition has shifted considerably more slowly (Dmitrijeva and Silovs, 2018).

According to Gildberg (1993), all the raw sources are by-products, inedible or edible, left during the processing of the core product. Bones, viscera, skin, and cut-offs left after filet cut are typically marine by-products that can be used for food processing. The by-products are fish heads and, sometimes, tails remaining after fish dismembering from cords used for smoking in the processing of the popular fish product canned "Sprats in oil." These by-products are typically used or refined for fish meal, although the lipid and protein content enables them to be considered as raw food materials. However, our analysis has shown that such goods may be utilized as a basis for the manufacture of human consumption products with value-added. "Sprats in oil" development cycle needs sprat smoking once the fish has been processed, cleaned, and threaded through the gills on rods. The fish is decapitated while smoking by slicing it off the rods (the fillet is removed from the head which resides on the rods). Filets are often utilized for canning, although normally smoked heads don't find it practical. Nevertheless, sprat heads composition 13 to 70% of the fish mass and comprise up to 55% of essential raw food resources-proteins and lipids (D'Amico et al., 2016).

There are many challenges in the extra production of the sprats smoked head, that avoid their use in food:

1. A huge amount of strong pieces (jaws, gills, skull bones,), which may be difficult to chew,
2. The limited size of heads that do not require edible sections to be distinguished from inedible pieces,
3. The powerful scent of smoke and a large concentration of smoke chemicals, which exceeds the appropriate concentration of food items,
4. A large sum of connective tissue and the unacceptable value of amino acids.

The study suggested the utility of the following technical operations aimed at overcoming the problems, requiring the sprats smoked heads to be treated as the raw food material:

1. Comprehensive grinding to maintain structural homogeneity. The biggest technical challenge is precise grinding of the raw resources of the fish. In industrial processing, excellently-configured heavy-speed cutters are utilized to precisely cut fish waste, or the pâté produced in a small-parameter cutter is to be treated as just a semi-finished product and might be moved via the emulsifying process. A quick cutter utilized during combination with the emulsifying is in many situations much safer than a costly high-speed cutter.
2. An elevated-value protein supplementation (vegetable, animal, or milk) to boost the amino acid ratings.
3. The usage of fillers to minimize the smoking substances

concentration to those approved for the usage in the food industry.

4. Treatment for softening of bone tissue with specific substances (if required). When selecting the volume of vinegar needed for the decalcification of bone tissue, attention might be given to the fact that, on the one hand, a large quantity of vinegar prevents the meat from spoilage and greatly softens the bone tissue of the cut heads of the fish; on the other hand, the continuous preservation of the pâté in the cans reduces the quantity of vinegar, allowing the food taste less-salty and acidic and also reduce the option that the pâté would lose its color due to long-time storage.
5. Usage of thickeners, textures, flavors, and spices to ensure the necessary parameters of organoleptic. Significantly cheaper mono-ingredient blends maybe supplement the TARI blend of food flavorings.
6. The option of the sterilization method relies greatly on the autoclave machinery and working standards at each company, and many other considerations. Taking into consideration the materials and microbiological examination of the pâté, this problem is to be discussed separately along with the business technologists.

Analyzing the manufacturing characteristics of by-products left after the development of "Sprats in Oil," brought us to the assumption that pâté is the most interesting food that can be produced from by-products of smoked sprats. Pâté is a finely shredded product which contains a considerable quantity of emulsifiers, fillers, and flavorings. Pâté has often been deemed a cheap commodity and so it has been found acceptable to use low-cost fish raw resources like sprats and its manufacturing by-products. One of the key problems yet to be addressed is the detection of substances that allow for prolonged shelf life, and preserving the required color and shape without modifying and retaining certain measures of quality. Now, the main problem is getting lighter colors and reducing the amount of E-substances in the formulation.

It ought to be noticed that the ultimate avoidance ought to be prevented, despite the product's goal niche utilizing costly ingredients. Nevertheless, maintaining the demand should not weaken the product's organoleptic functionality. Since smoking heads have a unique scent and taste of smoke, the alternative is to utilize products with a spicy and unique taste (black pepper, garlic, onion, etc.) instead of moderate spices. Fillers (starch, wheat cream, fibers, soy) ought to be utilized to lower the concentration of the substances produced during the smoking phase. To improve the water-binding ability of fish raw sources and moisture preservation, it is advised that phosphate blends be applied to the pâté along with salt during the first phase of processing. Water binders or fillers can be utilized to obtain the expected pâté texture. The analysis has enabled the development of pâté with suitable organoleptic properties. For all the features: scent, color, texture, aftertaste, flavor, and appearance (N is equal to 58), the pâté produced during the testing was graded from "average" to "good" during the assessment process.

A broad variety of items may be made with more complex flavors, focusing on the neutrally flavored pâté. The product range has been extended by incorporating flavors to achieve variants of the pâté such as:

1. "Mediterranean Herbs—with dried thyme and rosemary;
2. "Borodinsky" with cardamom and cumin;
3. *"Classical"*—it is the basic pâté colored utilizing the "Caramel-burnt sugar" coloring;
4. "Tomato" pâté which contains 20% of tomato paste and 80% of pâté; and
5. "Chilly" with red hot chilli pepper grains and flakes (utilizing the extract of liquid paprika, this pâté is colored).

## 8.4. FISHERY RESOURCES

The difficulty in the management of the biological resources of aquatic is owing to the several factors linked to technical factors, socioeconomic concerns, and environmental conditions and the complexities in the estimation of stochastic processes relevant to the populations being utilized. Following a quick overview of the latest technology at world level, the authors concentrate on the Italian and Mediterranean seas presenting details on global capturing of fishery stocks and economic implications of fishing operations, highlighting the need for a management strategy that ought to be resilient with uncertainty, appropriate for the fishing of multi-species and achieving ecosystem targets. In this regard, the writers present the key guidelines of the Code of Ethics for Responsible Fisheries and interventions highlighted in Council Regulation (EC) 1967/2006 and the Policy Guideline on Maritime Management (Gordon, 1953; Clark, 1980).Fishery sources are provided by communities of plant and animal species that are utilized by humans, particularly for the development of fishmeal for aquaculture goods, animal husbandry, and different types of decoration and crafts (Harrington et al., 2005). Aquatic species populations are preserved and created inside the marine ecosystem by complex abiotic and biotic interactions. Their growth and morphology are related to various natural factors operating in the environment (predation, productivity, hydrographic conditions, etc.) and human activities across the last 2 centuries, first, and foremost, fisheries. Thus related behavior impacting social structures and populations of organisms are induced by a single animal, man, in the natural environment, regulated by the biological evolution laws. This observation instantly illustrates how the biological aspects of natural environments interact with those of human populations in social and economic terms (Spagnolo and Cataudella, 2011).

Fishing may be described as catching or capturing natural species from the coasts, oceans, and inland waters (streams, rivers, lakes, etc.) and it is one of the ancient practices of mankind. Fishing is also a significant industrial operation that supplies seafood and thus generates jobs not just for those (fishermen) who catch the food (crustaceans, fish, mollusks) as well as for those who work along with the network from the building of vessels and supplies of equipment relevant to product marketing (Suski and Cooke, 2005).

Fisheries, fishery stocks, and system complexity linkages that involve ecological (about marine ecosystems and populations), technical (boats, equipment, generators, etc.), socioeconomic (sector workers) and, ultimately, administrative, and political problems. Organisms harvested by fisheries belong primarily to food sources whose first component is microscopic algae with a supply restricted to the light availability, less or more within the first two hundred meters of depth. Main phytoplankton production is linked not just to the accessibility of light as well as to the availability of nutrients (nitrogen, iron, phosphorus, etc.) and also to specific subordinate sources of energy (tidal flows, upwelling waves, etc.). Christensen and Pauly (1995) report that an average of 8% of the sea's primary global production funds global fisheries or, in other terms, this is the amount of primary production that is food for humans. This proportion rises in marine areas and falls in the open ocean (approximate 2%) and regions with the rising currents (between 24 and 35%), reflecting the improved efficiency of these natural environments in terms of man-made capital (Tacon, 1995).

The analysis of fishery resources means a significant jump in the difficulty of ecological research of the relationship and relation among the parties (living types along with chemical, physical, and climatic conditions, biological causes, etc.) that define an unrepeatable and special scenario. Studying correlations and interactions on specific space-time scales among natural events involve the convergence of multiple scientific disciplines. Social and Economics sciences play a significant part in matters relating to fisheries resource management.

## 8.5. FISHERY RESOURCE MANAGEMENT

Throughout the centuries and until the 19th century, a human might not even reasonably imagine that our behavior might contribute to the decline or even disappearance of the oppressed populations and even assumed that fish resources were inexhaustible because all that was required to obtain a greater volume was to expend more time and develop the means used in their recruitment, i.e., to raise the so-called "fishing efforts." While aquatic species habitats are sustainable, they are not inexhaustible. The population exploitable fraction is known as the "stock" and is categorized because of its dynamics connected to the recruitment of new people and their development, and also individual casualty levels (Pringle, 1985). The latter is also by definition and affecting the fisheries (fishing casualty) capturing of organisms. The growth in catches sometimes registered in certain areas is usually due to the growth in the population and improving fishing technologies, instead of the increased access to resources at sea. According to the recent figures, overall world output in 2012 amounts to 158 million tons, of which 67 million tons to aquaculture activity and 91 million tons relates to fishery captures (FAO 2014). The fishing activity retained at high rates can lead to one or more resources to break in the supply areas. More than 70% of world fish resources are utilized, some 20% are overexploited, and just 10% of resources remain underexploited by fisheries (Khan et al., 2012; FAO 2014).

An uncontrolled fishery is causing both economically and biologically overfishing of land. In bio-ecological terminology, resource overutilization may trigger not just the reduction of stocks of species to fishing operation (target species) but also several others of little or no economic benefit, but even ecologically significant, incidentally captured along with the first. The key goal of fisheries management has been to protect fish resources or, in other terms, to determine a degree of utilization that might offer a permanent or constant sustainable yield (MSY) to the overall yield by weight. At the same time, the key goal in terms of economics is to reach optimum stable economic efficiency, or MSE (King, 1995). Facing with the various fisheries problems, management priorities generally apply to the conservation and restoration of species and aquatic habitats in addition to the conservation of specific populations threatened by fishing, to sustain economic stability and save employment (Siar and Agbayani, 1994) (Figure 8.4).

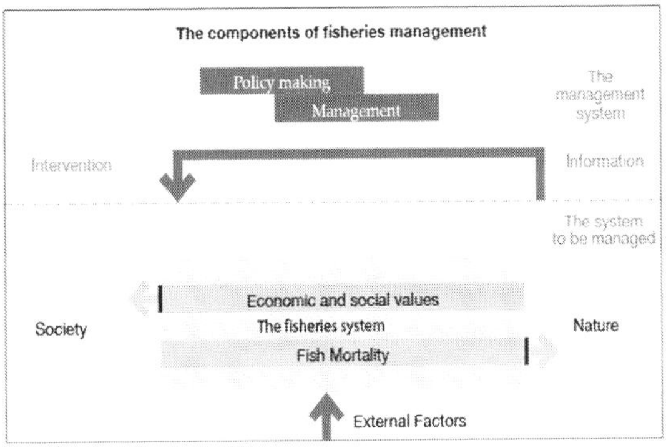


**Figure 8.4:** Fishery resource management systems.

*Source: http://regardssurlaterre.com/en/node/20017.*

## 8.6. ECOSYSTEM APPROACH TO FISHERIES (EAF)

Considering the fishing impacts on humans, habitats, and environments, the misuse of living resources is among the environmental problems that are currently being addressed on a global scale and as per ecological standards. Many aquatic ecosystems are threatened by fishing practices that include selective extraction for human food, economic gains, and production of the part of the natural resources (Maynou and Lleonart, 2003; Relini, 2003).

In 1995, the FAO Conference introduced the Code of conduct for Responsible Fisheries (FAO 1995) to foster long-standing balanced fisheries. For a holistic approach to marine resource management, in the widest sense of the term, seeks in the Code an overview of how to accomplish long-standing effective use of fishery capital in terms of trade, jobs, tourism, and food, and the socioeconomic and environmental well-being of communities worldwide. While some sections of the Code are focused on specific rules of international law, it is optional and includes criteria and guidelines for the management, protection, and growth of all fisheries. This also includes farming, manufacturing, and trade of fish, fishing initiative and strategies, fisheries, and aquaculture study (Maynou and Lleonart, 2003; Relini, 2003).

The Code has implemented by the FAO Conference of about 170 member states. Its implementation would be successful if the Member states turn their values and goals into specific policies and regulations (Bertrand et al., 2000; General Fisheries Commission for the Mediterranean (GFCM) 2011; FAO 2014) (Figure 8.5).

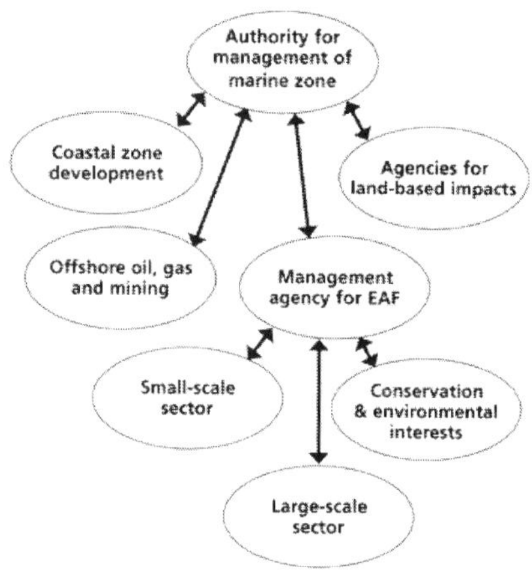


**Figure 8.5:** Paradigms of eco-system approach to fisheries.

*Source: http://www.fao.org/3/a0191e/A0191E09.htm.*

The international laws that have been implemented during the past 2 decades, along with the Code of Ethics for Responsible Fisheries, condemn the environment-based misuse of living resources; they emphasize the use of an ecosystem approach to fisheries (EAF). The concepts of an EAF are an expansion of the traditional standards for the creation of productive fisheries to include the whole ecosystem (Garcia et al., 2003). This suggests sustainable management not just of the industrial stocks but also of the entire ecological system that supports their development, along with the social and economic component (Spagnolo, 2006; Paterson et al., 2010).

The EAF predicts the implementation of various approaches and steps, from ecological to socioeconomic, to cope efficiently with complex circumstances involving a range of needs and demands. The goal of the EAF is to find a compromise between protection and sustainable usage of biodiversity. Particularly, it aims to fulfill the 3 sustainability dimensions, which are the social dimension (equity), the economic aspect (efficiency) and the ecological aspect (resource efficiency-re-productivity).

While the emergence of a Common Fishery Policy (CFP) dates back to the 1980s, the Common Fishery Policy's Green Book on the Future was only published in 2001 focusing on the Responsible Fisheries Code of Conduct. This chapter addresses the shortcomings of the fisheries policies that the European Union has adopted and the standards from which to build the aquaculture and fisheries reform project. In this regard, EC Regulation 2371/2002 actively encourages the effective use of resources from both a socioeconomic and an ecological viewpoint. Most specifically, as part of the CFP, EC Regulation 1967/2006, on conservation steps for the effective use of fishery stocks in the Mediterranean Sea, defines the Mediterranean Sea Action Plan (Spagnolo and Cataudella, 2011).

Under this Regulation, new guidelines for the conservation and safety of Mediterranean resources were developed with respect not only to commercial organisms but also to critical ecosystems and protected species. The Protocol contains the guidelines of the International Commission for the Conservation of Atlantic Tuna (ICCAT) and the Mediterranean General Fisheries Commission (GFCM), and the ecosystem management and species preservation initiatives stated in the Ecosystem Directive (92/43/EEC) and the Barcelona Convention (Maynou and Lleonart, 2003; FAO, 2014).

Finally, EC Regulation 1967/2006 accepts that the ecological, social, and economic features of Mediterranean fisheries allow the Society to carry out a clear management structure (EC Regulation 1967/2006). Fishing effort regulation is the key step in support of healthy Mediterranean fishing. Particularly, considering the variations in the allocation of living resources among the different geographical regions and the resulting various fishing customs and strategies, the regulations allow for the development of Regional and National Management Plans, linking the management of fishing activity with relevant technical interventions (EC Reg. 1967/2006).

The EC Regulation (1967/2006) authorized for more essential measures. Particularly, undersized people ought to be prevented from unnecessary captures. Therefore, such places in which juveniles (nursery regions) or adults (spawning regions) assemble must be secured. Fishing equipment which is too dangerous to the aquatic ecosystem or contributes to the reduction of some resources ought to be banned or supervised more strictly. To prevent more rises in mortality levels for juveniles and to decrease the number of discards of dead marine species by fishing vessels, it is important to improve the selectivity of the equipment currently being used and it improves the hook sizes for length, base-set nets and the net sizes for trawl nets (Carocci et al., 2009).The new regulations would allow marine reserves "no-take" and conservation initiatives to preserve the growing and fishing zones and the coastal ecosystem from harmful fishing impacts.

Marine reserves 'No-take' is assumed to be an alternative method of fisheries management focused primarily on general standards, like general public interest, ecological functions, and the requirement for socioeconomic risk and environmental insurance. In this context, starting in 1998, the Ministry of Agriculture, Food, and Forestry Policy (MiPAAF) declared, as result of the Fishery Resource Protection Program, the creation of "no-take areas" (Zone di Tutela Biologica, ZTB) where fishing is not permitted or controlled during the essential processes (spawning, recruiting, feeding, etc.) of the demersal stocks dispersed there, to provide for the renewal of stock and their effective utilization in the nearby regions.

These "no-take" zones would correlate to regions of financial interests that provide a key role in the lifecycle of one or more demersal organisms. In other terms, they will relate to places where organisms are concentrated for various biological purposes, and where conservation might improve the impact of management by closing them to fishing. While certain "no-take" areas along the Italian coast have been established, no data are yet available on their effects on fishery stocks (Bianchi and Fletcher, 2014).

The multi-species composition of Mediterranean fisheries, both in terms of organisms and gears of fishing, demands a unique approach capable of balancing and incorporating the various management initiatives to maintain stability in the fishing activities. In this case, the alternate establishment of a 'restricted season' in specific regions or capture limits and equipment restrictions in combination with the key legislation to limit fishing activity may prevent over-exploitation of marine resources (Garcia, 2003).

## 8.7. ALTERNATIVE INCOME-GENERATING FISHING ACTIVITIES

Overexploitation of aquatic resources continues to contribute to the reduction of several demersal reserves and the destruction of ecosystems, which is especially extreme in marine coastal areas where humans activities have been traditionally centered. The growing discharges of marine resources and waste, contaminations as well as tourism and industrial growth have also resulted in conclusive destruction of natural ecosystems with serious consequences for small-scale marine fisheries and consequent economic and social impact.

In contrast to the control of previously recorded fishing practices and the restoration and preservation of the aquatic ecosystem, in current years, the development of alternative initiatives have been introduced to move fishermen from fishing to tourism. The Ministerial Decree 293/99 generally controls "fishing tourism" that defines a fishing operation for educational and tourist services through which the local fishing society may boost its revenue (Rahman et al., 2012). Fishing tourism is the perfect opportunity to engage in local growth while simultaneously incorporating socioeconomic and cultural practices. The aim of tourist fishing is not just finding the sea's resources but also showing tourists how to value and observe it. This is the perfect place to understand the sea and fishing history, ponds, and coastal regions, and also to explore the old customs and occupations while still enjoying outstanding fresh cooked fish on the boat. All will benefit from the ancient history of fishing practices and Italian customs, in the same manner, such as in a natural park or museum. Additionally, the activity of fishing tourism extended to various seasons may lead to the seasonal adaptation of the flow of tourism, even the utilization of resource during the crucial time (Musuka and Chimba, 2014; Katikiro, 2016) (Figure 8.6).

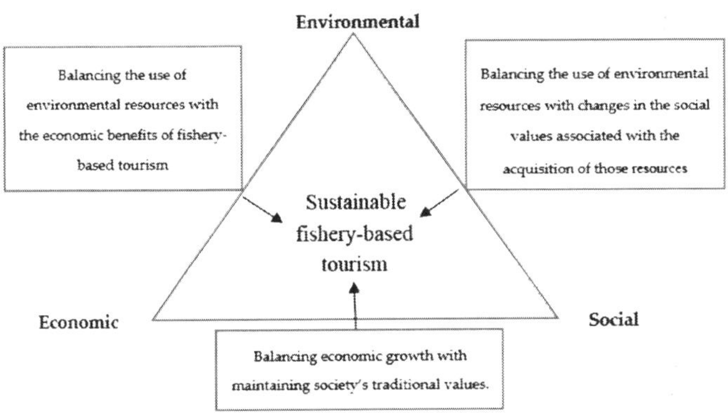


**Figure 8.6:** Sustainable fishery based eco-tourism has become popular in modern times.

*Source: https://www.mdpi.com/2079-9276/7/4/61/htm.*

With this practice, fishermen might reduce their overall effort of fishing and engage directly in the socioeconomic growth of regional. The incremental reduction in the key fishing operation created by the fishing tourism may ensure effective resource protection and preservation of habitat either directly via a commitment to conservation or indirectly through the provision of adequate income to community people to support their resources of wildlife as a source of revenue. An extra significant contribution that the fishing world may give is linked to tourism development based on conventional local goods and regional cuisine, and different events where the visitor plays an important part in the demonstration of fishing operations, even recommending the new services like environmental education, whale-dolphin watching and fishing tourism.

Human activity results in a reduction of biodiversity and species richness, and also a transition in population and society structure with a significant impact on ecological functions.

While the establishment of Marine Protected Areas (MPAs) along Italian shores is governed by the Ministry of the Environment for the preservation of ecosystems and the protection of the environment and not for fisheries like in other countries, the MPAs often reflect an effective and successful management policy for the effective usage of aquatic resources. Mediterranean MPAs offer vital services and products that sustain local economies and communities, including incentives for tourism and leisure, economic, and social growth, among other benefits. They might also encourage education and research for the climate. Additionally, MPAs preserve higher productivity and abundance in aquatic environments, offers protection for the survival of specific ecosystems and organisms, and also for endangered species that encourage them to regenerate, increase biodiversity and restock fishing waters by spilling out egg larvae, juveniles, and adults to replenish industrial stocks in neighboring fisheries.

## 8.8. STRATEGIES FOR THE MARINE ENVIRONMENT AND THE ECOSYSTEM APPROACH

At the same period, some of the above-mentioned management approaches to support and administrator consumer engagement. They often have a significant effect of embedding ecological priorities, considering resource-environment network sophistication (Backer et al., 2010).As mentioned in the model paper, it is widely recognized that stresses on natural coastal resources and demands for aquatic habitat services are still too strong and thus, irrespective of where their impacts are experienced, the need to reduce their effect on coastal waters is becoming even more urgent. The aquatic ecosystem, on the other side, is a valuable commodity which needs to be guarded, restored, and preserved, if necessary, to preserve biodiversity and protect the sustainability and wellbeing of oceans and seas that need to be healthy, productive, and clean. On 17 June 2008, to address such requirements, the Council of the European Union and the European Parliament published the Policy Framework Directive for the Marine Environment (EC Reg. 2008/56), which was also addressed while implementing laws in Italy (2001).

The Directive of the European Union is focused on a holistic strategy and aims at being the foundation of the European Union's future coastal policy. The Directive establishes the aim of a Good Environmental Status (GES) for member countries to meet by 2020 for their coastal waters. Member country should establish a coastal policy to meet GES, which should be accepted with neighboring Member countries and third countries (Bellido et al., 2011).Not just for the Marine Framework Directive, as well as for the European Union, preventing the lack of diversity is the basic objective today. Maintaining biodiversity is a basic necessity for coastal environments to achieve GES. The MPAs are an essential resource to ensure long-term biodiversity protection and ecological resources are preserved. To ensure biodiversity, the human aspect and the strong relation among the coastal marine environment and the economic, social, and cultural background of the local populations may not be overlooked. If the economic and social condition of their region is not considered, the ecological benefits of MPAs might become secondary (Apitz et al., 2006; Atkins et al., 2011).

Additionally, the framework aims to promote the fair usage of aquatic services and resources by an ecological strategy to human activity management, for example, a method that considers the population of human as an essential part of the habitats and the systems that control them (Thakur, 2014).At last, referring to the point of reference for this paper, the ecological concept is the guiding principle of the Marine Framework Directive (EC Regulation 2008/56), that was initially outlined in Twelve principles and may be divided into the below three main points: (1) Populations living in a region are accountable for the biodiversity that affects them. (2) There are three foundations of sustainability: economic, social-cultural, and environmental. (3) The control of the ecosystem needs an awareness of both the local and science dimensions (Leppänen and Backer, 2008).

## REFERENCES


1. AOAC, (2005). *Official methods of analysis of the Association of Official Analytical Chemists International* (18th edn., pp. 1–44), In: Association of Official Analytical Chemists, Washington (D.C.).
2. Apitz, S. E., Elliott, M., Fountain, M., & Galloway, T. S., (2006). European environmental management: Moving to an ecosystem approach. *Integrated Environmental Assessment and Management: An International Journal*, *2*(1), 80–85.
3. Arnesen, J. A., & Gildberg, A., (2002). Preparation and characterization of gelatin from the skin of harp seal (*Phoca groenlandica*). *Bioresource Technology*. *82*, 191–194
4. Atkins, J. P., Burdon, D., Elliott, M., & Gregory, A. J., (2011). Management of the marine environment: Integrating ecosystem services and societal benefits with the DPSIR framework in a systems approach. *Marine Pollution Bulletin*, *62*(2), 215–226.
5. Backer, H., & Leppänen, J. M., (2008). 1770721. The HELCOM system of a vision, strategic goals and ecological objectives: Implementing an ecosystem approach to the management of human activities in the Baltic Sea. *Aquatic Conservation*, *18*(3), 321–334.
6. Backer, H., Leppänen, J. M., Brusendorff, A. C., Forsius, K., Stankiewicz, M., Mehtonen, J., & Haaranen, T., (2010). HELCOM Baltic Sea action plan—a regional programme of measures for the marine environment based on the ecosystem approach. *Marine Pollution Bulletin*, *60*(5), 642–649.
7. Batista, I., Fradinho, P., & Silvestre, C., (2004). *Gelatin Extraction from Cape Hake and Blue Shark Skin* (pp. 1–30). 34th WEFTA meeting, Lubeck, Germany.
8. Baziwane, D., & He, Q., (2003). Gelatin: The paramount food additive. *Food Review International, 19*(4) 423–435.
9. Bellido, J. M., Santos, M. B., Pennino, M. G., Valeiras, X., & Pierce, G. J., (2011). Fishery discards and by catch: Solutions for an ecosystem approach to fisheries management? *Hydrobiologia*, *670*(1), 317.
10. Bergman, I., & Loxley, F., (1963). Two improved and simplified methods for spectrophotometric determination of hydroxyproline. *Analytical Chemistry, 35*(12) 65–75.
11. Bertrand, J. A., Gil, D. S. L., Papaconstantinou, C., Relini, G., & Souplet, A., (2000). An international bottom trawl survey in the Mediterranean:

The MEDITS programme. *Actes de Colloques IFREMER, 26*, 76–93.

12. Bills, T. D., Marking, L. L., & Mauck, W. L., (1981). Polychlorinated Biphenyl (Aroclor (R) 1254) Residues in Rainbow Trout: Effects on Sensitivity to Nine Fishery Chemicals. *North American Journal of Fisheries Management*, *1*(2), 200–203.
13. Bimbo, A. P., & Crowther, J. B., (1992). Fish meal and oil: Current uses. *Journal of the American Oil Chemists Society*, *69*(3), 221–227.
14. Bimbo, A. P., (1989). Fish oils: Past and present food uses. *Journal of the Americam Oil Chemists Society., 66*(12), 1717–1726.
15. Brezeski, M. M., (1987). Chitin and chitosan: Putting waste to good use. *INFOFISH International,* (5), 31–33.
16. Brine, C. J., & Austin, P. R., (1981). Chitin variability with species and method of preparation. *Comp. Biochem. Physiol*., 69(B), 283–286.
17. Bruno, S. F., Ekorong, F. J. A. A., Karkal, S. S., Cathrine, M. S. B., & Kudre, T. G., (2019). Green and innovative techniques for recovery of valuable compounds from seafood by-products and discards: A review. *Trends in Food Science* and *Technology*, *85*, 10–22.
18. BSI, (1975). *Methods for Sampling and Testing Gelatin (PhysicalandChemicalMethods)*(pp.1–30).BritishStandardInstitution, BS 75/1, London.
19. Campling, L., Havice, E., & McCall, H. P., (2012). The political economy and ecology of capture fisheries: Market dynamics, resource access and relations of exploitation and resistance. *Journal of Agrarian Change*, *12*(2, 3), 177–203.
20. Carocci, F., Bianchi, G., Eastwood, P., & Meaden, G. J., (2009). *Geographic Information Systems to Support the Ecosystem Approach to Fisheries: Status, Opportunities and Challenges* (Vol. 1, pp. 1–40). Doctoral dissertation.
21. Cataudella, S., & Spagnolo, M., (2011). The multidimensionality of fisheries: A brief introduction. In: Cataudella, S., & Spagnolo, M., (eds.), *The State of Italian Marine Fisheries and Aquaculture* (Vol. 1, pp 3–8) Ministry of Agricultural, Food and Forestry Policies (MiPAAF), Rome (Italy).
22. Chalamaiah, M., Hemalatha, R., & Jyothirmayi, T., (2012). Fish protein hydrolysates: Proximate composition, amino acid composition, antioxidant activities and applications: A review. *Food Chemistry*, *135*(4), 3020–3038.

23. Chimba, N. A. M. A. N. G. O. L. W. A., & Musuka, G., (2014). Impact of closed fishing season on the livelihood of fishers: A case of stratum I of Kafue fishery. *International Journal of Life Sciences Research, 2*(1), 49–62.

24. Cho, S. M., Kwak, K. S., Park, D. C., Gu, Y. S., Ji, G. I., & Jang, D. H., (2004). Processing optimization and functional properties of gelatin from shark (Isurus oxyrinchus) cartilage. *Food Hydrocolloid, 18*, 573–579.

25. Choi, S. S., & Regenstein, J. M., (2000). Physicochemical and sensory characteristics of fish gelatin. *Journal of Food Chemistry and Toxicology, 65*, 194–199.

26. Clark, C. W., (1980). Restricted access to common-property fishery resources: A game-theoretic analysis. In: *Dynamic Optimization and Mathematical Economics* (Vol. 1, pp. 117–132). Springer, Boston, MA.

27. Clausen, R., & Clark, B., (2005). The metabolic rift and marine ecology: An analysis of the ocean crisis within capitalist production. *Organization and Environment, 18*(4), 422–444.

28. Cooke, S. J., & Suski, C. D., (2005). Do we need species-specific guidelines for catch-and-release recreational angling to effectively conserve diverse fishery resources? *Biodiversity and Conservation, 14*(5), 1195–1209.

29. D'Amico, P., Armani, A., Gianfaldoni, D., & Guidi, A., (2016). New provisions for the labeling of fishery and aquaculture products: Difficulties in the implementation of regulation (EU) n. 1379/2013. *Marine Policy, 71*, 147–156.

30. Devi, L., Aparna, N., & Kalpana, K., (2013). Utilization of fish mince in formulation and development of pasta products. *International Food Research Journal, 20*(1), 219–224.

31. Diniz, F. M., & Martin A. M., (1997). Effect of the extent of enzymatic hydrolysis on the functional properties of shark protein hydrolysates. *Lebensmittel-Wissenschaft Und-Technology, 30*, 266–272.

32. EC Reg. 1543/2000, (2000). *Establishing a Community Framework for the Collection and Management of the Data Needed to Conduct the Common Fisheries Policy* (Vol. 1, pp.1–30). http://eur-lex.europa.eu/legal-content/EN/TXT/?uri=CELEX:32000R1543 (accessed on 30 May, 2020).

33. EC Reg. 1639/2001, (2001). *Establishing the Minimum and Extended*

*Community Programmes for the Collection of Data in the Fisheries Sector and Laying Down Detailed Rules for the Application of Council Regulation (EC)* (Vol. 1, pp. 1–22). http://eurlex.europa.eu/LexUriServ/LexUriServ.do?uri=OJ:L:2001:222:0053:0115:EN:PDF (accessed on 30 May, 2020).

34. EC Reg. 1967/2006, (2006). *Management measures for the sustainable exploitation of fishery resources in the Mediterranean Sea and Amending Regulation (EEC) 2847/1993 and Repealing Regulation (EC) 1626/1994* (pp.1–30). http://eur-lex.europa.eu/LexUriServ/LexUriServ.do?uri=OJ:L:2006:409:0009:0064:IT:PDF (accessed on 30 May, 2020).
35. EC Reg. 2008/56, (2006). *Establishing a Framework for Community Action in the field of Marine Environmental Policy (Marine Strategy Framework Directive*, Vol. 1, pp. 1–40). http://eur-lex.europa.eu/legal-content/EN/TXT/?uri=CELEX:32008L0056 (accessed on 30 May, 2020).
36. EC Reg. 2371/2002, (2002). On the conservation and sustainable exploitation of fisheries resources under the Common Fisheries Policy (Vol. 1, pp. 1–30), http://eur-lex.europa.eu/legal-content/en/ALL/?uri=CELEX:32002R2371 (accessed on 30 May, 2020).
37. Ellis, N., Pantus, F., & Pitcher, C. R., (2014). Scaling up experimental trawl impact results to fishery management scales—a Modeling approach for a "hot time." *Canadian Journal of Fisheries and Aquatic Sciences*, *71*(5), 733–746.
38. FAO, (1995). *Code of Conduct for Responsible Fisheries* (Vol. 1, pp. 1–33). Rome.
39. FAO, (2006). *Fish INFO Network Market Report, 1*, 1–30.
40. FAO, (2014). *The State of World Fisheries and Aquaculture* (Vol. 1, pp. 1–30). Food and Agriculture Organization of the United Nations, Rome.
41. FAO, (2014). *The State of World Fisheries and Aquaculture* (Vol. 1, pp. 10–44). Rome.
42. FAO, (2017). *Global Capture Production Database* (Vol. 1, pp. 1–29). Food and Agriculture Organization of the United Nations, Rome.
43. Farkas, V., (1990). Fungal cell walls: Their structure, biosynthesis and biotechnological aspects. *Acta Biotechnologica, 10*, 225–238
44. Fletcher, W. J., & Bianchi, G., (2014). The FAO-EAF toolbox: Making

the ecosystem approach accessible to all fisheries. *Ocean and Coastal Management*, *90*, 20–26.

45. García, C. B., Perez, D. A. N. I. E. L., Duarte, L. O., & Manjarres, L. U. I. S., (2008). Experimental results with a reducing device for juvenile fishes in a tropical shrimp fishery: Impact on the invertebrate by catch. *Pan-American Journal of Aquatic Sciences*, *3*(3), 275–281.
46. Garcia, S. M., (2003). *The Ecosystem Approach to Fisheries: Issues, Terminology, Principles, Institutional Foundations, Implementation and Outlook* (No. 443, pp. 1–30). Food & Agriculture Org.
47. Garcia, S. M., Zerbi, A., Aliaume, C., Do Chi, T., & Lasserre, G., (2003). *The Ecosystem Approach to Fisheries* (pp. 60–71). Issue, terminology, principles, institutional foundations, implementation and outlook. FAO Fisheries Technical Paper, N. 443, GFCM (General Fisheries Commission for the Mediterranean) 2011. Performance Review of the General Fisheries Commission for the Mediterranean and Black Sea [online]. http://www.fao.org/3/a-y4773e.pdf
48. Gehring, C. K., Gigliotti, J. C., Moritz, J. S., Tou, J. C., & Jaczynski, J., (2011). Functional and nutritional characteristics of proteins and lipids recovered by isoelectric processing of fish by-products and low-value fish: A review. *Food Chemistry*, *124*(2), 422–431.
49. Gildberg, A., (1993). Enzymic processing of marine raw materials. *Process Biochemistry*, *28*(1), 1–15.
50. Gimenez, B., Gomez-Guillen, M. C., & Montero, P., (2005a). The role of salt washing of fish skins in chemical and rheological properties of gelatin extracted. *Food Hydrocoll.*, *19*, 951–957.
51. Gordon, H. S., (1953). An economic approach to the optimum utilization of fishery resources. *Journal of the Fisheries Board of Canada*, *10*(7), 442–457.
52. Gordon, H. S., (1954). The economic theory of a common-property resource: The fishery. In: *Classic Papers in Natural Resource Economics* (Vol. 1, pp. 178–203). Palgrave Macmillan, London.
53. Grossman, S., & Bergman, M., (1992). *Process for the Production of Gelatin from Fish Skins*. US patent 5,093,474.
54. Habitat Directive (92/43/EEC), (2014). *On the Conservation of Natural Habitats and of Wild Fauna and Flora* (Vol.1, pp. 1–30). https://www.ecolex.org/details/legislation/council-directive-9243eec-on-the-conservation-of-natural-habitats-and-of-wild-fauna-and-flora-

lex-faoc034772/

55. Harrington, J. M., Myers, R. A., & Rosenberg, A. A., (2005). Wasted fishery resources: Discarded by-catch in the USA. *Fish and Fisheries*, *6*(4), 350–361.
56. Hasan, B., (2003). Fermentation of fish silage by using *Lactobacillus penthouses*. *Journal of Nature Indonesia, 6*, 11–15
57. Holzer, D., (1996). *Gelatin Production*. US Patent 5 484,888.
58. Jayathilakan, K., Sultana, K., Radhakrishna, K., & Bawa, A. S., (2012). Utilization of by-products and waste materials from meat, poultry and fish processing industries: A review. *Journal of Food Science and Technology*, *49*(3), 278–293.
59. Johnston-Banks, F. A., (1990). Gelatin. In: Harris, P., (ed.), *Food Gels*. (Vol.1, Chapter 7, pp. 233-389). Elsevier Applied Science, London.
60. Jones, N. R., (1977). Uses of gelatin in edible products. In: Ward, A. G., & Courts, A., (eds.), *The Science and Technology of Gelatin* (Vol. 1, pp 365–370). New York: Academic Press.
61. Jongiareonrak, A., Bennjkul, S., Visessanguan, W., & Tanaka, M., (2005). Skin gelatin from big eye snapper and brown strip red snapper: Chemical compositions and effect of microbial transglutaminase on gel properties. *Food Hydrocolloids, 20* 1216–1222.
62. Karim, A. A., & Bhat, R., (2009). Fish gelatins: Properties, challenges and prospects as alternative to mammalian gelatins. *Food Hydrocolloid, 23*, 563–576.
63. Katikiro, R. E., (2016). Improving alternative livelihood interventions in marine protected areas: A case study in Tanzania. *Marine Policy*, *70*, 22–29.
64. Khan, M. A., Alam, M. F., & Islam, K. J., (2012). The impact of co-management on household income and expenditure: An empirical analysis of common property fishery resource management in Bangladesh. *Ocean and Coastal Management*, *65*, 67–78.
65. King, M., & Faasili, U., (1999). Community-based management of subsistence fisheries in Samoa. *Fisheries Management and Ecology*, *6*(2), 133–144.
66. King, M., (1995). Fisheries biology, assessment and management. In: *Fishing News Books. Blackwell Science, 1*, 330–341.
67. Knorr, D., (1984). Use of chitinous polymers in food: A challenge for food research and development. *Food Technol., 38(*1), 85–97.

68. Koli, J. M., Basu, S., Nayak, B. B., Patange, S. B., Pagarkar, A. U., & Venkateshwarlu, G., (2011). Functional characteristics of gelatin extracted from skin and bone Tiger Toothed croaker (Otolithes ruber) and Pink pearch (Nemipterus japonicus). *Food and Bioproducts Processing, 1*, 1–40. doi: 10.1016/j.fbp.2011.08.001.
69. Koli, J. M., Sharangdher, S. T., Patange, S. B., Meter, S. Y., & Jain, A. R., (2015). Development of fish by-products by using fish and shellfish waste for up-liftment of socioeconomic status of fisher folk. *International Journal of Animal, Veterinary, Fishery and Allied Science*, *2*(1), 1–20.
70. Law no. 41, 17.02.1982, (2010). *Plan for the Rationalization and Development of Sea Fishing* (Vol. 1, pp. 5–22). http://www.normattiva.it/uri-res/N2Ls?urn:nir:stato:legge:1982-02-17;41@originale (accessed on 30 May, 2020).
71. Lleonart, J., & Maynou, F., (2003). Fish stock assessments in the Mediterranean: State of the art. *Scientia Marina, 67*(Suppl. 1), 37–49.
72. Lundqvist, J., De Fraiture, C., & Molden, D., (2008). *Saving Water: From Field to Fork: Curbing Losses and Wastage in the Food Chain* (Vol. 1, pp. 1–30*)*. Stockholm: Stockholm International Water Institute.
73. Mackie, I. M., (1983). New approaches in the use of fish proteins." In: Hudson, B. J. F., (ed.) *Developments in Food Proteins*, (Vol. 1, pp. 215–262). London: Applied Science Publishers Ltd.
74. Martin, A. M., (1999). A low-energy process for the conversion of fisheries waste biomass. *Renewable Energy*, *16*(1–4), 1102–1105.
75. McAllister, M. K., & Peterman, R. M., (1992). Experimental design in the management of fisheries: A review. *North American Journal of Fisheries Management*, *12*(1), 1–18.
76. Meyer, F. P., Schnick, R. A., Cumming, K. B., & Berger, B. L., (1976). Registration status of fishery chemicals, February, 1976. *The Progressive Fish-Culturist*, *38*(1), 3–7.
77. Miles, R. D., & Chapman, F. A., (2006). *The Benefits of Fish Meal in Aquaculture Diets* (Vol. 1, pp. 5–28). IFAS Extension, University of Florida.
78. Momot, W. T., & Gowing, H., (1977). Results of an experimental fishery on the crayfish Orconectes virilis. *Journal of the Fisheries Board of Canada*, *34*(11), 2041–2055.
79. Mora, C., Myers, R. A., Coll, M., Libralato, S., Pitcher, T. J., Sumaila,

R. U., & Worm, B., (2009). Management effectiveness of the world's marine fisheries. *PLoS Biology, 7*(6), 1235.

80. Mulye, V., Zofair, M., Baraiya, S., Solanki, K., Chudasama, J., & Temkar, S., (2014). Extrusion technology: Adding value to fish products. *INFOFISH International, 6/2014*, 1–30.
81. Muyonga, J. H., Cole, C. G. B., & Doudu, K. G., (2004). Characterization of acid soluble collagen from skin of young and adult Nile perch. *Journal Food Chemistry, 85*, 81–89.
82. Muyonga, J. H., Cole, C. G. B., & Duodu, K. G., (2004). Extraction and physicochemical characterization of nile-perch skin and bone gelatin. *Journal of Food Hydrocolloids, 18*, 581592.
83. Ninan, G., Zynudheen, A. A., & Joseph, J., (2011). Physico-chemical and textural properties of gelatins and water gel desserts prepared from the skin of freshwater carps. *Fish. Technol., 48*, 67–74
84. No, H. K., & Meyers, S. P., (1995). Preparation and characterization of chitin and chitosan: A review. *J. Aquat. Food Prod. Technol., 4*(2), 27–52.
85. Norland, R. E., (1990). Fish gelatin. In: Voight, M. N., & Botta, J. K., (eds.), *Advances in Fisheries Technology and Biotechnology for Increased Profitability* (Vol. 1, pp. 43–153). (Lancaster: T. Economic Publishing Co.).
86. Ockerman, H. W., & Hansen, C. L., (1988). Glue and gelatin. In: *Animal Byproduct Processing* (Vol. 1, pp. 1–30). Chichester, England: Ellis Horwood Ltd.
87. Otterer, M., (2002). Industrialization of fish silage. *Agropecuria, 1*, 190–200.
88. Paterson, B., Isaacs, M., Hara, M., Jarre, A., & Moloney, C. L., (2010). Transdisciplinary co-operation for an ecosystem approach to fisheries: A case study from the South African sardine fishery. *Marine Policy, 34*(4), 782–794.
89. Pauly, D., & Christensen, V., (1995). Primary production required to sustain global fisheries. *Nature, 374*, 255–257.
90. Pimentel, D., & Pimentel, M., (2003). Sustainability of meat-based and plant-based diets and the environment. *The American Journal of Clinical Nutrition, 78*(3), 660S–663S.
91. Pradhan, V. R., & Bedekar, A. N., (2002a). Study of biopolymer chitin: Costal and species variation. *Eco. Env. Cons., 8*(4), 341–343.

92. Pringle, J. D., (1985). The human factor in fishery resource management. *Canadian Journal of Fisheries and Aquatic Sciences*, *42*(2), 389–392.
93. Rahman, M., Rahman, M. M., Hasan, M. M., & Islam, M. R., (2012). Livelihood status and the potential of alternative income generating activities of fisher's community of Nijhumdwip under Hatiya Upazilla of Noakhali district in Bangladesh. *Bangladesh Res. Publ. J.*, *6*, 370–379.
94. Regenstein, J. M., (2004). Total utilization of fish. *Food Technology, 58*(3), 28–30.
95. Relini, G., (2003). Fishery and aquaculture relationship in the Mediterranean: Present and future. *Mediterranean Marine Science, 4*(2), 125–154.
96. Rönnbäck, P., (1999). The ecological basis for economic value of seafood production supported by mangrove ecosystems. *Ecological Economics*, *29*(2), 235–252.
97. Rustad, T., (2003). Utilization of marine by-products. *Electronic Journal of Environmental, Agricultural and Food Chemistry*, *2*(4), 458–463.
98. Salin, V., Atkins, J. A., & Salame, O., (2002). Value added in food manufacturing and retailing: A ratio analysis of major US states. *Journal of Food Distribution Research*, *33*(1), 136–150.
99. Sanmartín, E., Arboleya, J. C., Iloro, I., Escuredo, K., Elortza, F., & Moreno, F. J., (2012). Proteomic analysis of processing by-products from canned and fresh tuna: Identification of potentially functional food proteins. *Food Chemistry*, *134*(2), 1211–1219.
100. Sarabia, A. I., Gomez-Guillen, M. C., & Montero, P., (2000). The effect of added salts on the viscoelastic properties of fish skin gelatin. *J. Food Chem., 70*, 71–76.
101. Schnick, R. A., & Meyer, F. P., (1978). Registration of thirty-three fishery chemicals: Status of research and estimated costs of required contract studies. *Registration of Thirty-Three Fishery Chemicals: Status of Research and Estimated Costs of Required Contract Studies*, 86, 1–40.
102. Shahidi, F., & Botta, J. R., (2012). *Seafoods: Chemistry, Processing Technology and Quality* (Vol. 1, pp. 10–30). Springer Science & Business Media.
103. Shahidi, F., (1994). Proteins from Seafood Processing Discards. In:

Sikorski, Z. E., Sun, P. B., & Shahidi, F., (eds.), *Seafood Proteins* (Vol. 1, pp. 171–193). New York: Chapman and Hall.

104. Shahidi, F., (1995). Seafood processing by-products In: Shahidi, F., Botta, J. R., (eds.), *Seafood Chemistry, Processing Technology and Quality* (Vol. 1, pp. 320–334). New York, Kluwer Academic Publisher.
105. Shahidi, F., Arachchi, J. K. V., & Jeon, Y. J., (1999). Food applications of chitin and chitosans. *Trends in Food Science and Technology, 10*(2), 37–51.
106. Shanmugam, S. A., Gopalakannan, A., Jasmine, G. I., & Sugumar, G., (2001). Proteolytic *Lactobacillus spp.* in production of chitin and chitosan from shellfish waste. *Ind. J. Microbiol., 41*, 181–185.
107. Shirai, K., & Ramírez-Ramírez, J. C., (2011). 10 Utilization of fish processing by-products for bioactive compounds. *Fish Processing-Sustainability and New Opportunities, 1*, 230–236.
108. Silovs, M., & Dmitrijeva, O., (2018). Innovative technological process for emulgated pate production out of fish processing by-products. *Carpathian Journal of Food Science and Technology, 10*(4), 1–40.
109. Sneddon, C., (2007). Nature's materiality and the circuitous paths of accumulation: Dispossession of freshwater fisheries in Cambodia. *Antipode, 39*(1), 167–193.
110. Snedecor, G. W., & Cochran, W. G., (1967). *Statistical Methods* (6$^{th}$ edn., Vol. 1, pp. 580–593) Oxford and IBH-Publishing Co. New Delhi.
111. Songchotikunpan, P., Tattiyakul, J., & Supaphol, P., (2008). Extraction and electro spinning of gelatin from fish skin. *International J. Biol. Macromolecules, 42*, 247–255.
112. Spagnolo, M., (2006). Elements of economics and fisheries management (Vol. 1, pp. 200–279). FrancoAngeli, Milano.
113. Sperling, L. H., (1985). Introduction of physical polymer science (John Wiley, New York). *Indian Journal of Fundamental and Applied Life Sciences.* ISSN: 2231.
114. Sukumaran, K. K., (1987). Study on the fishery and biology of the mantis shrimp *Oratosquill*a nepa (latreille) of south kanara coast during 1979–83. *Indian Journal of Fisheries, 34*(3), 292–305.
115. Sundström, A., (2012). Corruption and regulatory compliance: Experimental findings from South African small-scale fisheries. *Marine Policy, 36*(6), 1255–1264.

116. Tacon, A. G., (1995). Feed ingredients for carnivorous fish species: Alternatives to fishmeal and other fishery resources. In: Reinertsen, H., & Haaland, H., (eds.), *Sustainable Fish Farming, 1,* 89–114.

117. Taskaya, L., & Jaczynski, J., (2009). Flocculation-enhanced protein recovery from fish processing by-products by isoelectric solubilization/ precipitation. *LWT-Food Science and Technology*, *42*(2), 570–575.

118. Thakur. A. K. (2014). *Handbook on Fisheries Statistics.* Department of Animal Husbandry, Dairying and Fisheries, Ministry of Agriculture (Vol.1, pp. 1–166).Government of India, New Delhi.

119. Thankappan, T. K., & Madhavan, P., (1985). A comparative study on chitosan with reference to yield and viscosity. *Paper Presented at the Symposium on Harvest and Postharvest Technology of Fish* (Vol. 1, pp. 591, 592), Society of Fisheries Technologists (India), Cochin.

120. Vásquez, J. A., (2008). Production, use and fate of Chilean brown seaweeds: Re-sources for a sustainable fishery. In: *Nineteenth International Seaweed Symposium* (Vol. 1, pp. 7–17). Springer, Dordrecht.

121. Venugopal, V., Shahidi, F., & Lee, T. C., (1995). Value-added products from underutilized fish species. *Critical Reviews in Food Science and Nutrition*, *35*(5), 431–453.

122. Vianna, M., & Almeida, T., (2005). Bony fish by catch in the southern Brazil pink shrimp (*Farfantepenaeus brasiliensis* and *F. paulensis*) fishery. *Brazilian Archives of Biology and Technology*, *48*(4), 611–623.

123. Wainewright, F. W., (1977). Physical tests for gelatin products. In: Ward, A. G., (ed.),*The Science Technology of Gelatins* (Vol. 1, pp. 508–531). Courts (Academic Press Inc., London).

124. Walstra, P., (2003). *Physical Chemistry of Foods*. Marcel Dekker, New York.

125. Ward, A. G., & Courts, A., (1977). *The Science Technology of Gelatins.* Academic Press Inc., London.

126. Wilson, J., Hayes, M., & Carney, B., (2011). Angiotensin-I-converting enzyme and prolyl endopeptidase inhibitory peptides from natural sources with a focus on marine processing by-products. *Food Chemistry*, *129*(2), 235–244.

127. Xu, L., & Geelen, D., (2018). Developing biostimulants from agro-food and industrial by-products. *Frontiers in Plant Science*, *9*, 1567.

128. Yang, H., Wang, Y., Jiang, M., Oh, J. H., Herring, J., & Zhou, P.,

(2007). 2-step optimization of the extraction and subsequent physical properties of channel catfish (*Ictalurus punctatus)* skin gelatin. *Journal of Food Science, 72*(4), 188–195.

129. Yasumatsu, K., Sawada, K., Moritaka, S., Misaki, M., Toda, J., Wada, T., & Ishii, K., (1972). Whipping and emulsifying properties of soybean product. *Agricultural and Biological Chemistry, 36*, 719–726.
130. Zar, H. J., (1999). *Biostatistical Analysis* (4th edn. Vol. 1, pp. 650–663). Dorling Kindersley (India) Pvt. Ltd., Delhi.
131. Zint, M., & Crook, A., (1998). A needs assessment of fisheries education materials for youth. *Fisheries*, *23*(10), 24–34.

# INDEX

B



C



D



E



F

## U



## V



## W